Notes On Cement Testing in Addition to Those Submitted May 24, 1900

You are holding a reproduction of an original work that is in the public domain in the United States of America, and possibly other countries. You may freely copy and distribute this work as no entity (individual or corporate) has a copyright on the body of the work. This book may contain prior copyright references, and library stamps (as most of these works were scanned from library copies). These have been scanned and retained as part of the historical artifact.

This book may have occasional imperfections such as missing or blurred pages, poor pictures, errant marks, etc. that were either part of the original artifact, or were introduced by the scanning process. We believe this work is culturally important, and despite the imperfections, have elected to bring it back into print as part of our continuing commitment to the preservation of printed works worldwide. We appreciate your understanding of the imperfections in the preservation process, and hope you enjoy this valuable book.

NOTES ON CEMENT TESTING

In addition to those submitted
May 24, 1900.

ALSO

ANSWERING REPORT

To one made March 5, 1901,

BY THE

ENGINEER OF THE DEPARTMENT OF HIGHWAYS,

Borough of Brooklyn.

On Violation of Specifications for Concrete Street Foundations.

REPORT

OF THE

COMMISSIONERS OF ACCOUNTS

OF

THE CITY OF NEW YORK.

MAY 20, 1901.

NOTES ON CEMENT TESTING

In addition to those submitted
May 24, 1900.

ALSO

ANSWERING REPORT
OF THE new york COMMISSIONERS OF ACCOUNTS

To one made March 5, (1901,)

BY THE
ENGINEER OF THE new york (city) DEPARTMENT OF HIGHWAYS, (1901.)

Borough of Brooklyn,

On Violation of Specifications for Concrete Street Foundations.

REPORT

TO THE

Hon. ROBERT A. VAN WYCK, Mayor,

MADE BY

JOHN C. HERTLE,
EDWARD OWEN,
Commissioners of Accounts of The City of New York,

MAY 20, 1901.

NEW YORK:
MARTIN B. BROWN CO., PRINTERS AND STATIONERS,
NOS. 49 TO 57 PARK PLACE.

1901.

Eng 809.01.3

Museum of Comparative Zoology

TABLE OF CONTENTS,

Showing Subdivision of Reports.

LETTER OF TRANSMISSION.

Letter of Transmission to Mayor ....pages	7 to	21
Findings .... "	15 to	19
Conclusions .... "	19 to	21

REPORT.

Report of Chief Engineer and Chemist............pages	23 to	74
Reasons calling for this report.......................page		27
Criticisms of Mr. McKenna's Letter..................pages	27 to	30
Criticisms of Letter of Engineer of Department of Highways, Borough of Brooklyn................ "	30 to	40
Scheme of Prof. Carmichael....................... "	45 to	47
Scheme of Mr. R. L. Humphrey .... "	47 to	48
Correspondence with a manufacturer................ "	49 to	53
Purpose of our Research .... "	54 to	55
Additional Notes on Cement Testing................ "	55 to	69
An Object Lesson................................ "	70 to	71
Our method of Analysis and Conclusions............. "	71 to	74

EXHIBITS.

" A."	Letter of Charles A. McKenna, which appeared in an Engineering Journal.....................pages	77 to	81
" B."	Editorial which appeared in same Journal ....page		82
" C."	Quotations from Letter of Engineer of Department of Highways, Borough of Brooklyn...........pages	83 to	90
" D."	Report of S. F. Peckham to Chief Engineer, referred to in answering report of N. P. Lewis.. "	91 to	93
" E."	Report of S. F. Peckham to Chief Engineer, referred to in answering report of N. P. Lewis.. "	94 to	96

Report of May 24, 1900 .... " 97 to 123

TABLES.

1.	Physical Tests of 34 samples of Cement............page		127
2.	Summary of Physical Tests...................... "		129
3.	Chemical Analyses .... "		131

LETTER OF TRANSMISSION

OF

COMMISSIONERS OF ACCOUNTS

TO

Hon. ROBERT A. VAN WYCK,

MAYOR.

LETTER OF TRANSMISSION.

OFFICE OF THE COMMISSIONERS OF ACCOUNTS,
STEWART BUILDING, NO. 280 BROADWAY,
NEW YORK, May 20, 1901.

SUBJECT: Notes on Cement Testing in addition to those submitted May 24, 1900,

also

Answering Report
of the Commissioners of Accounts
to one made March 5, 1901,
by the Engineer of the Department of Highways, Borough of Brooklyn, in reply to one made by the Commissioners of Accounts on violation of Specifications as to Concrete Street Foundations.

Hon. ROBERT A. VAN WYCK,
 Mayor :

DEAR SIR—As Commissioners of Accounts of the City of New York we beg to submit herewith for your consideration a report dated May 14, 1901, made to us by Mr. Otto H. Klein, our Chief Engineer, and Professor S. F. Peckham, Chemist in charge of our Laboratory.

In order to clearly demonstrate the necessity for this report and its importance to the city, it becomes necessary to call your Honor's attention to the following facts, viz.:

That on May 24, 1900, we submitted to you our printed report:

"Of a Comparison between Physical Tests and
"Chemical Analyses of 34 Samples of Portland and
"Rosendale Cements."

In said report we made use of the following language, viz.:

"On October 16, 1899, we made an additional report
"on violations of specifications for regulating, grading
"and paving contracts, recommending the substitution
"of Portland for Rosendale cement in concrete founda-
"tions, and in said report called attention to the fact
"that our conclusion was also shared by the Engineers
"of the Comptroller's office.   *   *   *

"Copies of said reports of May 4 and October 16,
"1899, were, by your Honor, transmitted to the De-
"partment of Highways, and shortly thereafter we were
"visited by several of the cement manufacturers, and
"the result of these interviews prompted us to send for
"the above-named 34 samples AND MAKE A CARE-
"FUL STUDY OF THE SUBJECT OF CEMENTS.

"Attached to this report will be found Tables Nos. 1
"and 2, showing the results we obtained from *physical*
"*tests*, and also Table No. 3, showing the results of the
"*chemical analyses* of the 34 samples of cements.
"(See pages 127, 129 and 131.)

"In these tabulated statements the samples are
"designated by a separate series of numbers for each
"test, for the purpose of not disclosing their identity.

"We realize the fact that up to the present time, so
"far as we have been able to discover, no *correspond-*

"*ence* has been observed between *physical tests* and
"*chemical analyses* of cements.

" This lack of correspondence appeared to us after
" the 34 samples had been analyzed, and, as a con-
" sequence, we made the *method* of analysis the sub-
" ject of investigation and developed a new process of
" analysis which, upon being applied to these thirty-four
" samples, the *physical tests* and *chemical analyses*
" showed corresponding results as to the quality of each
" sample of cement.

" The report, which we herewith submit, will, we
" believe, show beyond a doubt and demonstrate our
" finding that the *chemical analyses* of cements will
" always confirm the *physical tests*."

RESULT OF OUR PRINTED REPORT OF MAY 24, 1900.

The report of May 24, 1900, from which we have just quoted, has been the subject of much correspondence and criticism, both favorable and unfavorable, from Manufacturers, Scientific Men and Writers on this subject, and the demand for copies for this country and Europe has entirely exhausted our first edition.

Perhaps the most unfavorable criticism and which the Editor refused us the privilege of answering, appeared in an Engineering Journal of this City on August 18, 1900, in a letter signed by Dr. Charles F. McKenna, a commercial Chemist of this City, which letter, together with an editorial which appeared in said paper, will be found in Exhibits " A " and " B," pages 76 to 82.

VIOLATION OF CONTRACT SPECIFICATIONS IN BROOKLYN.

On November 10, 1900, about six months after the issuance of our Cement Report, we made an unfavorable report to your Honor on the violation of the specifications for laying a concrete foundation for an Asphalt Pavement on Watkins Street, Borough of Brooklyn, in which we reported that the physical tests made by us of the Rosendale cement used on this contract did not meet the very low requirements of the specifications of the Highway Department which form part of the contract between the Contractor and the City.

HIGHWAY ENGINEER'S ANSWERING REPORT, JAN. 4, 1901.

This report resulted in numerous answering reports between our Bureau and that of the Commissioner of Highways, the nature of which your Honor is informed, and finally we received, through your Honor from the Commissioner of Highways, a letter inclosing a report dated January 4, 1901, of 11 pages, made to him by his Engineer of Brooklyn, under whose charge the Watkins Street foundation was laid. So much of said letter as is necessary for the purpose of this report will be found in Exhibit "C," on pages 83 to 90.

COMMISSIONERS OF ACCOUNTS' ANSWERING REPORT.

In answer to the report of Mr. Lewis, Engineer of Highways, Borough of Brooklyn, on March 5, 1901, we submitted a report, of which the following is a copy, viz.:

NEW YORK, March 5, 1901.

"*Subject:* Answer in reference to Concrete Foundation on Watkins Street, East New York Avenue to New Lots Road. } Borough of Brooklyn.

" *Hon.* ROBERT A. VAN WYCK, *Mayor* :

" DEAR SIR—On December 13, 1900, we made a report to
" your Honor regarding the inferior cement used in the con-
" crete foundation for the asphalt pavement on Watkins
" Street, from East New York avenue to New Lots road,
" Borough of Brooklyn.

"On January 15 of this year, we received from your
" Honor a letter transmitting a communication dated Janu-
" ary 11, 1901, from the Hon. James P. Keating, Commis-
" sioner of Highways, with which was inclosed a report made
" to the Hon. Thomas R. Farrell, Deputy Commissioner of
" Highways, signed by N. P. Lewis, Engineer of Highways
" for the Borough of Brooklyn.

" Certain special examinations, the nature of which your
" Honor is aware, occupied our attention so closely that we
" could not, at the receipt of your letter, give the matter the
" attention we at that time thought it deserved.

" Said answering report of Engineer Lewis, of the High-
" way Department for the Borough of Brooklyn, consisted of
" about eleven pages, and referred principally to Chemical
" problems involved in the study of the composition of Ce-
" ments, as seen by Mr. Broadhurst, the young Chemist of
" the Department of Highways.

"This answering report was based largely upon the letter of Dr. Charles F. McKenna, published in the issue of the *Engineering Record* of August 18, 1900, said letter purporting to be a criticism of our methods of analysis of cement, as set forth in our report to your Honor, dated May 24, 1900.

"Inasmuch as we believe the criticism of Dr. McKenna to be neither just nor sincere, it is needless to say that we place no value on the statements contained in the report made by Engineer Lewis and his Chemist.

"With the transmittal to your Honor of our printed report on Cements, dated May 24, 1900, our investigation of the problems relating to the use and analysis of Cements, was not closed, but has been in progress ever since, and is still incomplete at this date.

"We deem it prudent, for the time being, not to discuss the matters involved in the report of Mr. Lewis, believing that the time is not far distant when we may be able to demonstrate, in a more practical manner than by a mere report to your Honor, the theoretical and practical correctness of our contentions.

"For the present, we beg to quote the following letter addressed to us by our Chemist in relation to the report of Engineer Lewis and his young Chemist:

"NEW YORK, January 26, 1901.

"*To the Commissioners of Accounts of The City of New York:*

"GENTLEMEN—I beg herewith to acknowledge the receipt of the report dated January 4, 1901, made to Hon. Thomas R. Farrell, Deputy Commissioner of Highways, Borough of Brooklyn, New York City, by Mr. N. P. Lewis, Engineer of Highways.

"This report has been given very careful consider-
"ation, especially as to the various statements and
"conclusions contained therein that pertain to the
"chemical analysis, and the chemistry of cements in
"general.

"To set forth in detail all the fallacies involved in
"these statements and conclusions would require more
"time and space than are now at my disposal. I
"would, however, briefly call attention to the follow-
"ing paragraph :

"'We, however, believe that the above
"'method of analysis to be a new, radical
"'and arbitrary one and at variance with
"'the usual practice, which is to employ
"'*concentrated* hydrochloric acid, and de-
"'termining all of the silica, alumina and
"'iron oxides, and lime so decomposed
"'from the silicates, aluminates, etc.
"'This method has given entire satis-
"'faction in the past, and we see no
"'reason for discarding it.'

"If improvements in analytical processes are con-
"demned because they are 'new, radical and arbi-
"trary,' progress in analytical chemistry would be im-
"possible.

"The history of analytical chemistry for the last
"hundred years consists of a perpetual succession of
"processes and methods which might be characterized
"as new, radical and arbitrary, in time becoming old
"and time-honored, but finally ceasing to give satisfac-
"tion, because new, radical and arbitrary processes and
"methods were *proved* to be better.

"That the use of concentrated hydrochloric acid in
"the analysis of cements has not 'given entire satisfac-

" ' tion in the past,' is proved by correspondence on file, " and also by the current literature of the engineering " profession for the last year, a literature with which " both Mr. Lewis and his Chemist, Mr. Broadhurst, " appear to be most surprisingly ignorant.

" Respectfully submitted,
" (Signed) " S. F. PECKHAM.

" In conclusion, we wish to state, notwithstanding " Engineer Lewis's contention that he found *some* Cement " which was used on this contract in question, which met " the requirements of the specifications, we found Cement, " which did *not* meet the requirements of the Department, " and as the City *pays* a liberal price for concrete street " foundations, we contend that the supervision and " inspection by the Department of Highways, should be " of such a character as not to admit *any material* which " will not meet the very moderate requirements exacted " by the specifications.

" That these requirements are moderate, is conclusively " shown by the fact that several brands of cement, now on " the market, exhibit an *excess* of tensile strength of nearly " 100 per cent.

" That defective Cement, in a considerable quantity, was " used in the concrete foundation on Watkins Street, is " demonstrated by the condition of the concrete itself, when " taken from the street in the presence of one of the " engineers of the Comptroller's office, weeks after it was " put down, and also by additional quantities, taken since by " our Engineer, and which are now in our possession.

" Respectfully,
" (Signed) " JOHN C. HERTLE,
EDWARD OWEN,
" *Commissioners of Accounts.*

Findings.

From the accompanying report of Messrs. Klein and Peckham, now submitted for your consideration, the following findings are submitted:

First—That Mr. Lewis, the Engineer of the Highway Department of the Borough of Brooklyn, while contending that wide differences are found between his results and our own, has by the results which he has stated, shown that those results are not only similar to our own, but that they confirm our own, when tabulated side by side.

(See pages 31 to 38.)

Second—That by the results exhibited by Mr. Lewis by the use of our method of analysis by his own Chemist upon similar specimens with our own, that similar results with our own are obtained; and further, that his contention that our methods are unique and arbitrary, is not sustained by the chemists, eminent in cement analysis, who have reported to the Committee on Cement Tests of the American Society of Civil Engineers.

(See pages 37, 38, 39, 45, 47 and 48.)

Third—That the entire reasoning and conclusions of Mr. Lewis loses sight of, or ignores, those constituents of Rosendale Cement that are inherent in the acknowledged imperfections of its manufacture, viz.:

The unburned and partially burned rock, which, while adding to the percentage of lime adds nothing to the percentage of cement.

(See pages 33 to 36.)

Fourth—That the results both physical and chemical, obtained by Mr. Lewis and ourselves, upon the Watkins Street Concrete, confirm each other, and when properly interpreted show that the COMMERCIAL ROSENDALE CEMENT used for that concrete is really quite a different material from that represented by the fallacious interpretation used by Mr. Lewis.

(See pages 37 to 40.)

Fifth—That the allegation of the Editor of the *Engineering Record*, which is nearly identical with that of Dr. McKenna, Mr. Lewis, and his Chemist, Mr. Broadhurst, that we apply the theoretical formula of pure Portland cements to natural (Rosendale) cements, is not sustained by any statement made in our report.

(See pages 40 and 41.)

Sixth—That concerning the contentions of the gentlemen named in the fifth finding, regarding our method of analysis, we refer chemists to the schemes of Messrs. Carmichael and Humphrey.

(See pages 45 to 48 and 71 to 74.)

Seventh—That concerning the criticisms of a practical cement manufacturer, we refer to our answer.

(See pages 49 to 53.)

Eighth—That, at the same time our chemist was working out his scheme of analysis which has provoked so much discussion, Professor Henry Carmichael, an acknowledged cement expert, formulated in entire independence, a scheme that is in most respects practically identical; also that the scheme of Mr. Humphrey is radically different from both Professors Carmichael and Peckham, and is not in accord with the generally recognized principles governing mineral analysis; also that the criticisms of our correspondent are found by us, upon analysis of his own raw material and products, to be theoretical rather than practical, and are not justified.

(See pages 61 to 69.)

Ninth—That legitimate criticism of our method of analysis lies, not so much in respect to the method itself, as in respect to the condition of the sample submitted to analysis; we on our part contending that inasmuch as the fineness of a cement is one of the qualities of a cement, and further, as our work has abundantly proved, that the solubility of cements depend upon their fineness, the samples of cement should be subjected to both Physical Tests and Chemical Analysis precisely as received at the Laboratory. We believe that our work has proved beyond any question that where the Physical Tests and Chemical Analyses are performed properly upon the samples as they are received, that the results correspond and confirm each other.

(See Tables 1, 2 and 3, pages 127 to 131.)

Tenth—That, in proof of the inadequacy of Humphrey's method of analysis, we prepared a mixture of pulverized fire brick and lime, which on analysis by his method, exhibited the silica, alumina and iron and lime, in the proper proportions for a first-class Portland Cement, while our own method exhibited it just as it was, viz., a mixture of insoluble material and lime, and not cement at all.

(See pages 44, 70 and 71.)

Eleventh—That our contention, that Rosendale cements are greatly inferior to Portland Cements for the making of concrete for street foundations is fully sustained, and that therefore Rosendale cements should not be used for that purpose.

CONCLUSIONS.

The Engineer of the Department of Highways for the Borough of Brooklyn, responsible for the concrete foundation on Watkins Street, Borough of Brooklyn, which was by us reported to your Honor, defends himself on the strength of the letter of Dr. McKenna and the editorial which appeared in the same journal on a previous date, and also upon the information which he received from the Chemist in charge of the Laboratory of the Highway Department.

The necessity therefore devolved upon us of demonstrating the fallacy of the contentions, upon which the Engineer of the Department of Highways for the Borough of Brooklyn bases his conclusions, in order that the City may in the future, not suffer from the construction of improper concrete street foundations made of cement which is allowed to be used by public officials of The City of New York under erroneous conclusions based upon false premises.

We finally wish to present to your Honor and for the benefit of those officials who have not seen our first report, on a

> "Comparison between Physical Tests and Chemical
> "Analyses of 34 samples of Portland and Rosendale
> "Cements,"

the following statement:

First—That complaints had reached this office that, in some instances where asphalt pavements had failed, that the failure had been without doubt due primarily to the lack of the necessary strength in the concrete foundation.

Second—That these complaints led to a general investigation of the cements used in these foundations and to the relation of Physical Tests to Chemical Composition.

Third—That in the course of this investigation it was found that the Portland Cements of good quality of both Foreign and Domestic brands, tested from 50 to 100 per cent. above the requirements of the Department of Highways for Portland Cements, and that they were more than 200 per cent. above the requirements of the same Department for Rosendale Cements.

Fourth—That the average Portland Cements, of a good quality, in the market exhibit a tensile strength of from 4.5 to 6 times that required in the specifications of the Department of Highways for Rosendale Cements in Street Foundations.

Fifth—That while Portland Cements for use in street foundations are mixed with THREE PARTS of sand and from six to seven parts of broken stone, Rosendale Cements are mixed with TWO PARTS of sand and four parts of broken stone only, thus making the cost of a street foundation laid of Portland cement nearly the same as that made of Rosendale cement.

Sixth—That Mr. Lewis, the Engineer of the Department of Highways in the Borough of Brooklyn, in testing 79 samples of " Commercial Rosendale Cement," used in street foundations in the Borough of Brooklyn, found that they averaged just above the ridiculously low requirements of the specifications of the Department of Highways, and that they ranged from below test to 40 per cent. above, and that more than 7 per cent. of the 79 samples were below test.

Seventh—That notwithstanding this fact, and that Portland Cements which are from $4\frac{1}{2}$ to 6 times as strong as the requirements of the specifications of the Department of Highways for Street Foundations, and from the further fact that there is very little difference in the cost by the use of Portland Cement, Mr. Lewis, the Engineer of the Department of Highways for the Borough of Brooklyn, appears to be satisfied that "Commercial Rosendale Cement" is a proper material from which to construct Concrete Foundations for asphalt Street Pavements.

Eighth—That we do not agree with him, in this respect, nor do we believe that Rosendale Cement is a suitable material from which to construct any Street foundation, whether the surface be of asphalt, brick or any other material, when concrete four times as strong can be had for the same or nearly the same outlay. (See pages 32 to 40).

    Respectfully submitted,

                JOHN C. HERTLE,
                EDWARD OWEN,
        The Commissioners of Accounts.

REPORT

OF

OTTO H. KLEIN

AND

STEPHEN F. PECKHAM

TO THE

COMMISSIONERS OF ACCOUNTS.

REPORT OF

OTTO H. KLEIN and STEPHEN F. PECKHAM.

ENGINEERING BUREAU.
OFFICE OF THE COMMISSIONERS OF ACCOUNTS

NEW YORK, May 14, 1901.

Subject: Notes on Cement Testing, in addition to those submitted on May 21 and 24, 1900.

Hon. JOHN C. HERTLE *and* EDWARD OWEN, *Commissioners of Accounts:*

GENTLEMEN—We had the pleasure of bringing before you in a Report, dated May 21, 1900, the results of a research made by us that had been in progress for more than a year, in the Physical and Chemical Laboratories of The Commissioners of Accounts of the City of New York, upon the Relations between Physical Tests and Chemical Analyses of Cement.

CRITICISM OF FORMER REPORT.

As might be expected, this Report has been made the subject of various criticisms, which are found to fall into three classes, viz.:

First—Those which are wholly commendatory.

Second—Those which are adverse, but given in courtesy.

Third—Those which are of such a character as to be deserving only of silent contempt, except for certain considerations other than their merits.

First Class of Criticism.

As an illustration of the first class, we offer the following from one of the leading chemists of the country:

" I am most deeply grateful for your thoughtful
" kindness in sending me copies of your very important
" reports. I had occasionally been called upon to an-
" alyze cements, according to the book method, but
" always wondered what it availed my client. I have
" often studied comparatively the tests of engineers
" along with those of chemists, but always gave it up as
" an insoluble puzzle. It is truly refreshing to have the
" riddle solved as you seem to have done it."

The second class will be discussed farther on.

Third Class of Criticism.

To the third class belongs a letter addressed to the editor of *The Engineering Record*, and which appeared in the issue of that Journal of August 18, 1900, over the signature of Charles F. McKenna, a commercial chemist.

In order not to make the body of the report too voluminous, we have attached a copy of said letter to this report and marked Exhibit "A," which will be found on page 77.

REASONS CALLING FOR THIS REPORT.

We desire to briefly reply to this letter on the following grounds, viz.:

First—For the reason that we were refused that privilege by the Editor of the paper at the time the letter and editorial article appeared.

Second—For the reason that this letter has been quoted and used by an official of the City of New York, in correspondence with our office, and in criticism of the work of this office, using it as an argument upon which to base what we believe to be wholly erroneous conclusions.

Our Criticisms of Mr. McKenna's Letter.

Paragraphs *a* and *b* of this letter are devoted to certain ethical considerations and to the expression of certain personal opinions, unprofessional, and therefore wholly out of place in a discussion, where the argument stands or falls on the merits of a process of chemical analysis.

Paragraph *c* is disingenuous and untrue.

Paragraph *d* contains a number of assertions which are left without proof either by reference or discussion. That a method of analysis and the conclusions derived from it or by means of it, is absurd, because it is unique, may be the opinion of the author of the letter, but such an opinion finds little support from the history of chemistry.

In paragraph *e*, his discussion of the term "soluble silica," wherein he begs the question is simply hypercriticism; it will be replied to farther on.

In paragraph *f* he asserts, "he uses five grammes throughout the analysis without subdividing." * * * This statement is an absolute untruth, as any one can see by reading our paper.

In paragraph *g* the author of this letter shows himself so fresh in the matter of mineral analysis and such a blind adherent of what he considers established methods, that he condemns any attempt to test their accuracy.

Why should any understanding be reached in reference to a matter of which we possess no knowledge?

Either the solution of cement in hydrochloric acid of varying strength gives constant and corresponding results or it does not.

Either the manner of solution has an effect upon the determination of the constituents of the cement or it has not.

Moreover, the percentages of lime were not totals, as he asserts, but they were, like the percentages of soluble silica, the amount passing into solution in the acid of varying strength.

The only astonishing thing about it is that Dr. McKenna should be willing to use a method of analysis, that furnished erroneous results and condemn a man for attempting to ascertain the facts and formulate a method capable of giving results more accurate.

In paragraph *h* he does not make his meaning clear; but he appears to think that insufficient burning of a mixture of which limestone or marl is one of the constituents, will not result in the presence of carbonate of lime in the burned material, but that carbonate of lime when present in a cement must be purposely added as an adulterant.

In paragraph *i* he says:
"the carbon he finds he ascribes in all cases to unburnt "coal."

The word coal cannot be found in our report, nor can one word be found there which refers to the source of the carbon. The carbon is present all the same, in some cements, whatever its source may be.

In paragraph *k* he assumes that we were not doing exactly what we had been doing for some months. This work with standard acid is not identical in results with the gravimetric method described by us and was not mentioned by us. In course of time we expect to get to it.

We quite agree with Dr. McKenna in his conclusions stated in paragraph *l*. No one, so far as we know, besides Dr. McKenna has wasted any gray matter over "active index."

In paragraph *m*, after asserting that we make an attempt which any one who reads our Report will find is exactly what we do not attempt, he finds it all "shockingly absurd." We quite agree that this paragraph is "shockingly absurd," if so mild a term will properly characterize it.

We would like to believe that in writing this letter Dr. McKenna was sincere, but we find it impossible to accept the apology with which it closes. To do so would force us to question his intelligence. Moreover, we believe if he had been sincere his letter would have been carefully truthful and would have exhibited the manners of a professional gentleman, which it does not.

Mr. N. P. Lewis' Letter.

The official correspondence referred to in our "Second Reason," on page 27 of this report, is a letter dated January 4, 1901, addressed by Mr. N. P. Lewis, Engineer of the Department of Highways for the Borough of Brooklyn, to his Deputy Commissioner, which finally reached us, for our attention, through his Honor the Mayor, on January 15, 1901.

A very careful examination of this letter at the time it was submitted led us to defer any extended reply to it until certain investigations then in progress had been completed.

As this letter involves a discussion of points of vital interest in the cement problem, we quote from it at length, in order that the subject matter may be more clearly brought before you.

So much of Engineer Lewis' letter as is necessary for the purpose of this report is hereto attached and marked Exhibit "C," see pages 83 to 90.

Our Criticism of Letter of Eng. Lewis.

A general criticism of this letter of Mr. Lewis and of the letter of Dr. McKenna, which is quoted, lies in the atmosphere of oracular profundity and occultism that pervades them. Thoughts too big for utterance seem to lie behind the obscure hints that are thrown out, but which are not fully stated. What is really meant by the authors can only be conjectured. Careful comparisons of what is plainly expressed reveals the fact that Mr. Lewis' letter, including that of his chemist, Mr. Broadhurst, is as "shockingly" insincere as the letter of Dr. McKenna.

PHYSICAL TESTS OF COMMERCIAL ROSENDALE CEMENT.

Since Mr. Lewis has seen fit to make public, in this letter to the Department of Highways, the fact that the cement under discussion is Commercial Rosendale cement, we proceed to use such evidence as is on record in this office and such additional evidence as Mr. Lewis has furnished in his letter without regard to any private interests that may be involved.

Four samples of this brand, numbered 45, 78, 83a and 142, of cement have been subjected to physical tests in our Physical Laboratory, with the following results, viz. :

	45	78	83a	142
One day's neat............	48 pounds.	22 pounds.	30 pounds.	34 pounds.
Seven days' neat	87 "	65 "	67 "	58 "
Seven days' mortar.........	172 "	10 "	17 "	24 "

Mr. Lewis makes use of the following language, viz. :

" As to lack of uniformity in these 79 lots, I will call
" your attention to the fact that on one day neat tests
" one lot showed over 70 pounds, two lots showed be-
" tween 60 and 70 pounds, 70 lots showed between 50
" and 60 pounds, and 6 lots showed less than 50 pounds,
" the lowest being 46 pounds and the average being 53
" pounds.

" On the seven-day neat tests one lot showed over
" 140 pounds, two lots showed between 130 and 140
" pounds, 10 lots showed between 120 and 130 pounds,
" 16 lots showed between 110 and 120 pounds,
" one lot showed under 110 pounds, it being 106
" pounds, while the average was 120 pounds. * * *
" The average results obtained when mixed with two
" parts of standard sand were 7 days, 55 pounds."

He does not give the details of the seven-day mortar tests.

Chemical Analyses of Commercial Rosendale Cement.

There are in this office records of more or less complete analyses of four samples of Commercial Rosendale Cement, and Mr. Lewis' report furnishes us with two more. They are as follows:

NUMBERS.	COMMISSIONERS OF ACCOUNTS' SAMPLES.				LEWIS' SAMPLES.	
	90a.	90b.	90c.	279.	Oct. 23.	Nov. 25.
Insoluble in 10 per cent. HCl.........	9.73	8.82	23.98	16.738	9.486	13.277
Carbon.............................	0.72	Trace.	0.54	0.182	0.985	1.022
Magnesia...........................	Trace.	......	......	2.300	2.843	2.861
Sulphuric oxide.....................	1.73	......	......	Trace.	0.760	0.759
Matter volatile at red heat...........	8.65	......	19.15	13.17	13.369	14.972
	20.83	......	43.67	32.39	27.44	32.89
Soluble silica.......................	15.24	17.95	7.92	15.06	14.74	13.13
Alumina and iron oxide..............	6.48	8.42	3.03	6.97	5.73	5.44
Calcium oxide (lime)................	53.76	56.14	42.86	45.54	52.10	48.52
	75.48	82.52	53.81	67.57	72.57	67.09

Discussion of these Results of Analysis.

These results show, that Commercial Rosendale Cement is extremely variable in its composition, and that the variation in the amount insoluble in 10 per cent. HCl is nearly 100 per cent.

This material consists of the constituents of the cement rock, that for the most part has not been sufficiently burned to cause the lime and silica and alumina to combine into soluble forms, and also of iron that has been burned sufficiently to render it insoluble in dilute acid, all of which is not cement, and performs no other office in the mixture than so much sand.

The carbon adds, by whatever the amount may be, to this matter that is not cement.

The magnesia and sulphuric oxide in these samples are not in excessive amount; in, fact they are low in all of them.

The next item of importance is the matter volatile at a red heat, which varies by more than 100 per cent. It indicates directly the amount of carbonic acid remaining in the cement from insufficient burning, and indirectly either the impossibility of making a uniform quality of cement from natural rock or else the lack of skill, or care, or both, in the manufacture of this particular brand of cement. The work that we have done shows that in these six specimens under consideration from 8 to 16 per cent. of them is carbonic acid, averaging at least 12 per cent. and representing in round numbers 27.25 per cent. of unburned carbonate of lime, or nearly, if not quite one-third of the limestone or cement rock.

That these conclusions are sound is demonstrated by the physical tests.

No. 90a on a seven-day mortar test gave 175 pounds, equal to 388 per cent of the tensile strength required.

No. 90c gave 27 pounds, exactly 60 per cent. of the tensile strength required. Mr. Lewis does not mention the tensile strength of the cement of November 25, but it was doubtless up to the requirements.

Nos. 90a and 90c were taken off of works in Manhattan and Brooklyn.

The strong cement No. 90a shows but 8.65 per cent. volatile at a red heat, which indicates about 8 per cent. carbonic acid, equivalent to 18.2 of carbonate of lime, containing 10.2 per cent. of lime. 53.76 per cent. less 10.2 per cent. = 43.56 per cent. of lime in combination with 6.48 of alumina and iron oxide and 15.24 per cent. of soluble silica to form cement, equal to 65.28 per cent. of the amount taken.

When this cement is mixed with twice the amount of sand, the mixture represents 65 parts of cement to 235 parts of sand, or roughly 1:3.5. (See physical tests page 32, No. 45.)

No. 90c. shows 19.15 per cent. volatile at a red heat, which indicates about 16 per cent. carbonic acid, equivalent to 36.32 per cent. of carbonate of lime, containing 20.32 per cent. of lime. 42.86 per cent. less 20.32 per cent. = 22.54 per cent. of lime in combination with 3.03 per cent. of alumina and iron oxide and 7.92 per cent. of soluble silica to form cement, equal to 33.49 per cent. of the amount taken. Where one part of this cement is mixed with two parts of sand, the mixture represents 33.49 parts of cement to 266.5 of sand, or roughly 1:8. (See physical tests, page 32, No. 83a.)

Mr. Lewis' cement of November 25, shows 14.972 per cent. of matter volatile at a red heat, which indicates about 12 per cent. of carbonic acid, equivalent to 27.27 per cent. of carbonate of lime, containing 25.27 per cent. of lime; 48.52 per cent. less 15.27 = 33.25 per cent. of lime in combination with 5.44 per cent. of alumina and iron oxide and 13.13 per cent. of soluble silica to form cement equal to 51.82 per cent. of the amount taken. Where one part of this cement is mixed with two parts of sand, the mixture represents 51.82 parts to 248.18 parts, or roughly 1:5.

A certain bag or barrel contains a certain finely ground substance called Rosendale cement. All of the cement there really is, in either, is represented by the silica, alumina and iron oxide and lime that are so combined that they will form an hydraulic mixture. There may be a small quantity of plaster of Paris that adds something to the cohesive strength, but it is not hydraulic cement. This cement is soluble in very dilute acid—even a 2 per cent. solution of glacial acetic acid dissolves it, as also other very weak solutions of organic acids.

All that there is in the bag or barrel, that is not cement as above described, is worth less for purposes of making mortar than a like amount of clean sand. This material may be pulverized, unburned cement rock, or the same partially burned, the ashes of the fuel used, the pulverized fine fragments of the fuel used (which we call carbon, but which Mr. Lewis calls "Insoluble matter volatile on ignition"), iron oxide which has not combined to form cement, magnesia, sand, etc.

Where these materials are present in moderate amount, as in No. 90a, the mortar resulting from its use is nearly as strong as Portland cement mortar. Where they are very considerable in amount, as in the sample of November 25, the resulting mortar is not worth more than half as much as Portland cement mortar. Such cement as is represented by 90c is good for nothing.

THE WATKINS STREET CEMENT.

This brand of cement was used on Watkins Street, Brooklyn, and has been analyzed by ourselves and by Messrs. Lewis and Broadhurst. Our own analysis is No. 279. There are no differences between the three samples, as shown by the analyses, that are of any practical value.

They all three belong to the class above mentioned, where the materials above mentioned, that are not cement, are found in very considerable amounts, resulting in a mortar, that at best, is not worth more than one-half as much as Portland cement mortar.

The Watkins Street Concrete.

The concrete that has been produced from the use of this cement has also been investigated by ourselves and by Messrs. Lewis and Broadhurst.

	\multicolumn{4}{c}{Separation of Samples of Concrete according to}			
	\multicolumn{2}{c}{Commissioners of Accounts.}		\multicolumn{2}{c}{Lewis.}	
Numbers.	280A.	280B.	H59.	H60.
Total weight of samples...	595 grams.	832 grams.	1,059.09 grams.	3,794.05 grams.
Weight of stone............	360 "	547 "	668.50 "	2,563.95 "
Per cent. of stone..........	60 per cent.	65 per cent.	63.12 per cent.	67.59 per cent.
Weight of mortar..........	235 grams.	285 grams.	390.59 grams.	1,230.10 grams.
Per cent. of mortar........	40 per cent.	35 per cent.	36.88 per cent	32.41 per cent.

These results show an extreme difference in the mixture of stone and mortar in the four samples of 7.5 per cent. There is about the same difference between Lewis' samples as between our own. As Mr. Lewis' results were computed on larger samples they may be more nearly correct, but the differences do not appear to us to be material.

The mortar was air dried and sifted from the stone. Our samples effervesced freely when introduced into 10 per cent. dilute hydrochloric acid, and we have no doubt the others did, but Mr. Broadhurst does not mention it.

The matter insoluble in the acid consisted of:

	COMMISSIONERS OF ACCOUNTS.		LEWIS.	
NUMBERS.	280A	280B	H59	H60
Insoluble Matter.				
Sand and other mineral matter	69.60	71.35	71.18	71.34
Carbon	.52	.63	.70	.83
Total	70.12	71.98	71.88	72.17
Soluble Matter.				
Silica	3.68	3.38	3.62	3.46
Alumina and Iron Oxides	5.85	3.56	2.85	2.81
Lime	12.44	12.59	13.78	13.46
Magnesia	Trace	Trace	Trace	Trace
Sulphuric Oxide	"	"	"	"
	92.09	91.51	92.13	91.90

These results, like those determining the proportions of the concrete, are as nearly alike as could be expected, and confirm each other.

DISCUSSION OF THE RESULTS OF ANALYSIS OF THE CONCRETE.

We do not, however, agree with the conclusions that Messrs. Lewis and Broadhurst have drawn from them. The statement is made in Mr. Lewis' letter, paragraph 12, that, " this Commercial Rosendale cement contains approximately 30 per cent. of so-called " Inert matter " (Matter Insoluble in 10 per cent. Dilute hydrochloric acid) we do not know of any Rosendale cement which contains 15 per cent. of matter corresponding to this figure."

It is fortunate that Mr. Lewis took so much pains to state exactly what he means and also that he means by matter insoluble in 10 per cent. dilute hydrochloric acid exactly the matter referred to in Exhibit "E" of this report on pages 94 to 96, when we based certain calculations on an assumption of a Rosendale cement containing 15 per cent. of matter insoluble in 10 per cent. dilute hydrochloric acid. Some of them contain a great deal more than 15 per cent. of such matter, but none that we have examined contained 30 per cent.

Referring to our table, on page 33, we find that of these six samples of Commercial Rosendale cement, three of which are from our former report, and two from Mr. Lewis' letter, the matter insoluble in 10 per cent. dilute hydrochloric acid is 9.73, 8.82, 23.98, 16.74, 9.49 and 13.28, of which the average is 13.67.

Mr. Lewis shows in Exhibit "C," paragraph 11, that a mortar consisting of one part cement to two parts of sand by volume, contains by weight 71.33 per cent. of sand, and 28.67 per cent. of cement. 13.67 per cent. of 28.67 = 3.86 pounds of inert matter in the amount of this average Rosendale cement contained in 100 pounds of mortar. 3.86 added to 71.33 = 75.19 which leaves 24.81 of cement, provided the cement *is* cement; but we have already shown (on page 34) that this cement containing at least 12 per cent. of carbonic acid is at least one-quarter carbonate of lime, amounting to 6.2 per cent. or 6.2 pounds which should be added to the 75.19 pounds = 81.39 pounds of inert matter to 18.61 pounds of cement in 100 pounds of dry mortar.

Estimating Commercial Rosendale Cement, as cement, it is easily shown that the mortar used on Watkins Street filled the requirements of the specifications. Estimating this cement at what it really is, a mixture of cement, carbonate of lime and other inert matter, it is not surprising that one month after the concrete was laid the mortar could be pulverized between the fingers and removed from the stone with the thumb nail.

Instead of being 75.19 of inert matter to 24.81 parts cement, the mortar made from these Commercial Rosendale cements, consists of 81.39 parts to 18.61 parts; in other words, there is only about three-fourths as much cement in the mortar as the specifications call for.

FURTHER CONSIDERATION OF CRITICISM OF FORMER REPORT.

Returning to the criticisms of our report of May 24, 1900, as shown below, we beg to consider them under "Four Specific Allegations," viz. :

First Allegation.

"That we carried on our investigations of Rosen-
" dale cements along the same lines as our investiga-
" tions of Portland cements, and that such position is
" untenable.

In reply to this half truth, we quote from our former report, as follows :

"It could not be expected that a cement made by burning a natural lime rock containing clay and silica, would be found, on chemical analysis, to conform in its composition to any theoretical formula. The fact that natural cements do not, could not be more forcibly demonstrated than by an inspection of the accompanying table. * * * No other results could be expected from careless and rapid burning of stone in large lumps in kilns or stacks than great lack of uniformity in the result and consequent uncertainty as to the value of the product."

While we believe this statement to be a sufficient reply to this puerile and insincere criticism, we remark further, that we have never observed any claim that there was one chemical compound that formed the hydraulic portion of Portland cements and another and different compound that made Rosendale cements hydraulic. We think any such claim cannot be sustained, and, therefore, that any direct or indirect attempt to advocate the use of Rosendale cements as equivalent to Portland cements on such grounds is both injudicious and unfair.

The constant soliciting of consideration for Rosendale cements that is not extended to Portland cements, is in itself a confession of weakness and is, in our judgment, in the long run bound to react unfavorably.

We have never claimed that Rosendale cements "contain ingredients in such proportions as to make a theoretically perfect cement."

Second Allegation.

"That our methods of analysis are new and unique, radical and arbitrary, and at variance with the usual practice," etc.

If improvements in analytical processes are condemned because they are new, etc., then progress in analytical chemistry would be impossible. The history of analytical chemistry for more than a century consists of a perpetual succession of processes and methods which might be characterized as new, radical and arbitrary, which in time became old and time-honored, but which finally ceased to give satisfaction, because other new, radical and arbitrary processes and methods were proved to be better.

Third Allegation.

"That the use of concentrated hydrochloric acid has given entire satisfaction in the past."

That this statement is wholly incorrect is proved by correspondence on file in our office and also by the current literature of the engineering profession for the past year, which we shall quote farther on in this report, a literature with which we are surprised to find both Mr. Lewis and his chemist, Mr. Broadhurst, apparently unfamiliar.

Fourth Allegation.

Contained in paragraph 8 of Mr. Lewis' letter.

" It would appear as though the chemist had first in-
" tended to examine the concrete as a whole, but he sub-
" sequently confined his investigations to the mortar."

This statement is not clear, as the stone was nothing but fragments of old bluestone curbing; why should a chemist examine them?

The chemist did exactly what he intended to do, viz. :

To determine the proportions of stone and mortar, and this he found to be practically the same as did Mr. Lewis.

If Mr. Lewis had been sincere in his comparison, he would have given our percentages and shown them to be practically the same as his own. He would also have used the words "inert matter"; in one sense instead of two.

Mr. Lewis defines his use of the words "inert matter" to be the matter insoluble in 10 per cent. hydrochloric acid and then proceeds to use the term to signify the "inactive constituents" of the cement, which includes everything in the sample that is not cement.

If the words are used as he defines them, it is perfectly correct to consider that an average sample of Rosendale cement will contain 15 per cent. of such inert matter.

If he uses the words as equivalent to inactive constituents then 30 per cent is the proper figure.

It is very apparent from his definition of the term and use of it, that his reasoning in paragraph 12 is insincere and that he knew he was not constructing his argument upon premises laid down by the Commissioners of Accounts, in their Report of May 24, 1900, as he alleges in paragraph 13.

The Second Class of Criticisms.

As concerning the second class of criticisms and before proceeding to discuss the merits of the method of analysis which was described in our former Report, we wish to call attention, first, to the

> "'Progress Report of the Special Committee on the "'proper manipulation of tests of cement,' made to "the American Society of Civil Engineers, and pub-"lished in the Proceedings of that Society, Vol. XXVI, "No. 4, April, 1900."

This report consists of the various replies, to a long list of questions, offered by many persons, and submitted by the chairman of the committee, Prof. G. F. Swain, without comment.

SCHEME OF PROF. HENRY CARMICHAEL.

In reply to Question 5,

What elements of compounds should be determined?

Professor Henry Carmichael of Boston (who is an acknowledged authority as a cement expert), says:

> "Hydraulic cement consists of a double silicate of lime "and alumina (including iron oxide), which is readily "soluble in dilute hydrochloric acid, leaving little or no "insoluble residue. In addition to the soluble silica "and the oxide of calcium, aluminum and iron, good "cement contains traces of the oxides of magnesium, "sodium and potassium, together with traces of carbon-"ates, sulphates, chlorides and combined water, and "finally minute amounts of insoluble sand or cinder."

In reply to Question 6,

"*What do you consider the best methods of determining these compounds with sufficient accuracy?*"

Prof. Carmichael continues,

"The sample is ground fine in an agate mortar. One gram is carefully weighed out in a shallow porcelain dish and well covered with a 3 per cent. solution of hydrochloric acid. After several hours the cement should completely dissolve in this acid with the exception of a small amount of sand, mostly black cinder, from the fuel employed in making the cement. The residue, if any, is filtered off and determined. The clear solution is evaporated to dryness on a water bath in a flat dish. Hydrochloric acid is poured over the dry residue, and the acid is then evaporated. Add a few drops of same acid, again drive off acid. Moisten residue again with same acid and boil up with pure water. The silica is rendered insoluble by the above operation and can be filtered off and weighed. The silica which thus dissolves in the dilute acid, and is in turn rendered insoluble, is the silica which is available in the setting of the cement. The filtrate from silica is boiled with a few drops of nitric acid, and pure ammonia is then added, which precipitates the oxides of iron and aluminum. With the ammonia is added also ammonium chloride in sufficient quantity to retain the lime in solution. After boiling for some time, the oxides of iron and aluminum are filtered off, and after drying are ignited and weighed."

Here follow directions for separating the iron and aluminum:

"To the filtrate from iron and aluminum oxides is added a slight excess of ammonium oxalate, whereby the lime is precipitated as oxalate which is filtered off, ignited at a dull red heat in a platinum crucible and weighed as carbonate."

His scheme offers further details for the determination of the ingredients that he says are found in good cements in traces; for the determination of water and carbonic acid by ignition; and for the determination of free lime by titration.

This scheme closely resembles our own, while differing from it in several important respects. It was worked out without any knowledge whatever on the part of either party concerning the other. A proper comparison will show that many of the puerile criticisms indulged by Dr. McKenna concerning our paper, apply with equal force to the scheme of Prof. Carmichael.

SCHEME OF R. L. HUMPHREY.

Following the scheme of Prof. Carmichael is another scheme by R. L. Humphrey, who writes as follows:

" One-half gram of the finely pulverized sample
" dried at 100° C., is thoroughly mixed with four or five
" times its weight of sodium carbonate, and fused in a
" platinum crucible until CO_2 no longer escapes; the
" crucible and its contents is placed in a beaker, and
" twenty or thirty times its quantity of water, and about
" 10 c.c. of dilute HCl is added; when complete solu-
" tion is effected, it is transferred to a casserole and
" placed on a water bath, and evaporated to dryness
" several times. The mass is taken up with dilute HCl
" and water, heated for a short time and filtered, wash-
" ing the residue on the filter thoroughly with hot water.
" The filter is dried, ignited and weighed. This weight
" (less ash) gives the amount SiO_2."

" The filtrate is brought to boiling and ammonium
" hydrate added in slight excess, the boiling is continued
" until the odor of ammonia is no longer perceptible.
" Filter and wash. Redissolve in hot dilute HCl, again
" precipitate with ammonia and filter through the pre-
" vious filter and wash with boiling water. The precipi-
" tate dried, ignited and weighed, less ash, gives the
" amount of Al_2O_3 and Fe_2O_3."

Then follows a method of separating iron from alumina:

" The filtrate from the iron and alumina is heated to
" boiling, and boiling ammonium oxalate is added until
" a precipitate is no longer formed. After boiling for a
" few minutes, it is set aside for a short time; when
" the precipitate has settled perfectly, decant the clear
" liquid through a filter, wash by decantation, dissolve
" the precipitate in hot dilute HCl, using as small a
" quantity as possible to effect a complete solution, heat
" to boiling and add ammonia, heat on a water bath for
" a few minutes; when the solution clears, filter through
" the previous filter, wash thoroughly with hot water.
" Dry the precipitate, ignite to constant weight, and
" weigh as CaO", or determine the lime volumetrically
" by titration with potassium permanganate."

He then determines the ingredients occurring in small proportion. He determines SO_3 in a separate portion after removing the silica.

Correspondence.

We wish to call attention secondly, to correspondence which we have lately held with a prominent manufacturer of cement, in which he indulged in the following criticisms of our paper.

(*a*) "Experiments we have made show that the "solubility of commercial Portland cements in dilute "acid depends greatly upon the fineness of grinding."

(*b*) "We have found no cements which if ground to "extreme fineness in an agate mortar, show more than "a fraction of one per cent. of insoluble matter."

(*c*) "It seems to me therefore, that your separation "of the components of cement into active and inactive "constituents is not well grounded."

(*d*) "I think it probable that the active index would "be somewhat reduced if the cement is dissolved as "completely as possible."

(*e*) "Since even the monosilicate of lime, wallasto- "nite, is readily decomposed by acid, it is evident that "the residue should contain practically no lime, and "would consist of a minute amount of uncombined clay.

(*f*) "It is impossible that this should reach more "than a fraction of one per cent. in a good cement.

(*g*) "Would say further that the use of sufficiently "dilute acid and fine grinding will give a clear solution "without any separation of gelatinous silica.

(*h*) "In my opinion no conclusion can be drawn "from the amount of lime soluble in water.

(*i*) "This would also depend greatly on the degree "of pulverization. (*ii*) It is generally held, as you "know, that all cements are decomposed by water into "calcium hydrate and a hydrated monosilicate. (*iii*) If "this is true, the action of water would be progressive, "and prolonged action of sufficient water would dis- "solve out all the lime. (*iiii*) In fact, Le Chatelier "found this to be the case."

Our Answer to Manufacturer's Criticism.

To which we replied:
The first sentence of your letter furnished a key to the whole matter. You say:

" That experiments we have made show that the
" solubility of commercial Portland cements in acid de-
" pends greatly upon the fineness of grinding."

Believing that this fact as stated by you has been proved beyond any question, we insist that every sample of cement shall be analyzed in exactly the condition in which it is brought to the laboratory; that is, that the specimen shall be neither *dried nor pulverized*, nor in any manner treated in such a way as to either lessen or increase the differences that exist between the samples as they are brought upon the works or are subjected to physical tests.

We have not yet found a weighable amount of lime in any of the residues from dilute acid that we have examined.

We think that if you read our paper carefully you will see that no conclusions are drawn from the amount of lime soluble in water. It is, however, an observed fact that all of the good cements contain about 5 per cent. of such soluble lime.

If you would advocate the uniform grinding of samples of cement to an impalpable powder, in an agate mortar, in order that they may be more completely dissolved and brought into solution, we insist that we do not agree with you.

We believe that cement of proper fineness for use is soluble in 10 per cent. hydrochloric acid without gelatinization, and that any matter not so soluble, contained in commercial cement, is not cement at all, and is, and should be classed as, 'inert matter.'"

To which he replied,

(*j*) "your separation of the constituents of cement "into 'active and inactive constituents' by the action "of 10 per cent. hydrochloric acid on the commercial "cements, appears to me to be without foundation.

(*k*) "The varying amounts of insoluble matter "obtained by you on treating the same cement with "acid of varying strengths and in various ways appear "to show that the amount of insoluble matter depends "upon the method employed, rather than upon the "chemical character of the cement analyzed.

(*l*) "I have found that most commercial Portland "cements, if ground to great fineness, give scarcely any "insoluble residue on treatment with sufficient quantity "of 5 per cent. acid.

(*m*) "The high percentage of insoluble matter "obtained by you simply results from the compara-"tively coarse grinding of the cement.

(*n*) "You certainly will not contend that the com-"position of the coarser particles is materially different "from that of the fine ones, or that the chemical char-"acter of the cement is changed by grinding the coarse "particles to uniform fineness with the rest.

(*o*) "Your choice of 10 per cent. acid and method "of mixing appear to me to be wholly arbitrary, and the "conclusions drawn from the amount of insoluble mat-"ter obtained to be quite unjustified.

(*p*) "Since coarsely ground cement gives a consider-"able residue when treated with dilute acid, while finely "ground cement gives practically none, and since this "residue consists chiefly of silica, and contains, as "stated by you, practically no lime, it appears to me "evident that this insoluble matter results chiefly from "local separation of silica contained in the coarse "particles.

(*q*) "The lime and other constituents contained in "these particles are, however, dissolved, and are in- "cluded by you in the group of active constituents.

(*r*) "The injustice of this is apparent.

(*s*) "If the silica of the coarse particles is inactive, "the lime must be so also.

(*t*) "It is undoubtedly correct to submit commercial "samples of cement to *physical tests* as they are re- "ceived, without grinding.

(*u*) "To submit these samples to chemical analysis, "however, without bringing them into homogeneous "condition by grinding, is certain to lead to erroneous "conclusions.

* * * * * * * * * *

(*v*) "In burning, however, a disturbing factor enters, "and this is the ash of the coal dust used as fuel.

(*w*) "This ash adds at least two per cent. to the "silica, alumina and iron oxide of the product.

(*x*) "It is, however, brought into combination with "the lime of the charge sufficiently to become wholly "soluble in acid, but not uniformly enough to allow "the lime of the raw material to be raised to a corre- "sponding extent.

(*y*) "I believe fully that the best Portland cements "are thoroughly homogeneous in character, and that "the excess of silica, alumina and iron over that called "for by the formula is due to the ash of the fuel and to "the general practice of carrying the lime a little below "the maximum in order to offset possible fluctuations "in the mixture."

We asked our correspondent to send us a sample of the fuel ash.

He replied:

(*z*) "It will be impossible to send you a sample of the coal ash to which you refer, as this melts in with the clinker with which it comes in contact in the rotary kilns.

(*za*) "The amount of fuel used is about 150 pounds to the barrel of cement.

(*zb*) "The ash of this fuel is about 8 per cent., and if all absorbed by the clinker would add about 3 per cent of silica, iron and alumina to the latter."

THE PURPOSE OF OUR RESEARCH.

We wish to discuss in association these methods of analysis and this correspondence, all of which comes from unquestioned authority and is strictly professional, all of which represent a purpose that is in fundamental opposition to our own purpose, together with the letters and reports previously cited.

The purpose of the research described in our former paper was to ascertain, whether or no any correspondence could be established between the results of a chemical analysis and the physical tests of cement. It was not to defend or defame any brand of cements, to contrast any one cement with another, or to contrast American cements with foreign cements, or one class of foreign cements with another class of the same, although some very unexpected results obtained from the research led incidentally to such comparisons.

This object was kept clearly in view from first to last and while the different brands of cement examined represented in nearly every case nothing but a name; still, the different samples were designated wholly by numbers that were purposely made different in the two laboratories.

The cements first analyzed were brought to the office from various works in progress in the City, and they gave results that were extremely unsatisfactory apparently for various reasons that we do not care to take the time and place to discuss here.

Keeping the object of our research in view, we solicited samples of freshly ground cement from a number of manufacturers and dealers, and we still found that when these fresh, and in most instances, high class cements were analyzed by the methods described in the books, that no correspondence could be observed between the physical tests and chemical analyses.

Believing that Nature could not contradict herself, we were convinced that there must be some defect in the question put to nature, in the chemical analyses; in other words, that the chemical analysis was not properly conducted. This led us to conclude that the purpose of an analysis was not satisfied when an, in some respects, inadequate method of *ultimate* analysis was followed.

ADDITIONAL NOTES ON CEMENT TESTING.

THE COMPOSITION OF CEMENTS.

For, no cement, either Portland or Rosendale or slag, on the market, consists of chemically pure hydraulic cement. Assuming the correctness of the researches that led up to and include those of the Messrs. Newberry, we have a right to further assume that hydraulic cement consists of tri-calcic silicate and ferro-aluminate in the proportions indicated by the equation.

$$\frac{CaO}{2.8 \times \text{Silica} + (1.1 \times \text{Alumina \& ferric oxide})} = 1$$

A commercial Portland cement therefore consists of the above-named compound plus, not a trace, but an *unavoidable percentage* of the ashes of the coal employed as fuel,

> Also of overburned clinker,
> " under burned clinker,
> " uncombined clay,
> " Magnesia,
> " Sulphuric oxide,
> " Alkalies, and
> " a small percentage of water and CO_2.

In the case of Rosendale cements, there is in addition to these an unavoidable percentage of both overburned and underburned cement rock, particularly the latter, together with more or less minute fragments of the fuel used that are ground up with the cement and appear as carbon.

We assume that these impurities are unavoidable, because from the nature of the case no cement can be manufactured without them and no cement that we have examined has been entirely free from them, and we believe that we have examined some of the best cements now made in the world.

The amount of carbon ranged in those we have examined from zero to nearly 2 per cent. This carbon is not a proper constituent particularly of a Portland cement. No good cement contained more than a trace of it. It is, and must be from the nature of the case, a source of weakness, especially when it is in the form of an oily soot from imperfect combustion of oil used as fuel.

These methods of analysis do not offer any opportunity for the determination of this impurity, and it does not make it anything but carbon to call it " Insoluble matter volatile on ignition."

Also of Plaster of Paris, introduced purposely to influence the setting of the cement. When moderate in amount this is not an injurious ingredient.

Also, a percentage of quartz sand. Stillman's and Humphrey's methods of analysis offer no opportunity for the determination of this ingredient.

Also a percentage of underburned rock or clinker which results in the presence of carbonate of lime, which is inert, having no cementing properties. That it is possible to manufacture a Rosendale cement, containing but little of it is shown by analysis No. 90a, page 33.

Stillman's and Humphrey's methods furnish no opportunity for the determination of this ingredient.

Cements made from marl are liable to contain Glauconite, which is a mineral containing a comparatively large amount of Alkalies. Cements made from slag are also liable to contain alkalies. We have no reason to suppose, however, that alkalies exist in any of the samples that we have examined in any injurious amount, and they were therefore not determined.

The Question Put to Nature.

The question therefore to be put to Nature is, how much Hydraulic Cement does this mixture contain as it is brought to the Laboratory? And what else does it contain that it should not?

How the Question is Answered.

In order to answer this question, it may first be determined how much of the sample of the cement is volatile at a bright red heat. We have found by repeated examples that an ordinary Portland cement of good quality will give about 4 per cent. or less, of a matter volatile at a red heat; with Rosendale cements the amount is much greater. A number of tests showed that this amount was not materially increased by the use of a blast lamp. In a few instances the increase amounted to a few tenths per cent. but in a majority of instances it was not appreciable. This loss at a red heat was found in a majority of instances to very nearly correspond with the amount of carbonic acid determined directly, but in those instances where this loss was large, amounting to more than 10 per cent, the difference between these two factors increased as the loss increased, showing that in those cases where this loss is large, an increasing proportion of such loss is water. If more satisfactory, the percentage of carbonic acid can be directly determined, but even when it is so determined, the matter volatile at a red heat should always be ascertained.

The Sample should be Analyzed as Received.

The problem of analysis is simply an ordinary problem in mineral analysis. The physical tests have been made upon the sample as brought to the Laboratory and in such a manner as to exhibit the greatest possible differences between this sample and all other samples. The samples were neither dried nor pulverized, because either drying or pulverizing, or both, would make another and different sample to be tested from the one submitted. Our correspondent admits this, and if he did not everybody else does; yet, referring to his sentence marked (u) he claims that to submit these samples to chemical analysis, without bringing them into homogenous condition by grinding, is certain to lead to erroneous conclusions. Both schemes of analysis that we have quoted require that the sample shall be ground, and one of them that it shall be dried. We have found the loss that good cements sustain when dried at 100° C., is not of practical importance, with bad cements it is otherwise.

We believe that right here is the keynote to the controversy, and that in discussing any subject the disputants must be agreed as to their subject or they may argue forever and both be correct in reaching diametrically opposite conclusions.

In this case the two chemists and our correspondent are agreed that the samples should be pulverized to an impalpable powder, and we, on the contrary, believe that as soon as a sample is pulverized it is converted into a new sample. This our correspondent practically admits. (See sentences $a, b, d, f, g, l, m, p, q, s, y$.)

We claim that the sample should be analyzed as received and thus made to exhibit the greatest possible differences, as is the case in the physical tests. If the samples are pulverized and thus brought uniformly into solution, certain differences are lost, and it is not strange that the physical tests and chemical analyses do not correspond.

The same result follows when the samples are fused with sodium carbonate and when the sample is dissolved in concentrated HCl. In both instances the distinction between the silica that is combined into tri-calcic-ferro-alumino-silicate or cement, and the silica that exists as sand or fuel ash in unburned clay is lost.

Active and Inactive Constituents.

In our former report we did not base any distinction of "active" and "inactive" constituents on the action of 10 per cent. HCl (*o, j*). We base those distinctions on the researches of the Messrs. Newberry and we insist that if those researches are conclusive, as we believe that they are, our conclusions follow from the nature of the case. There was nothing arbitrary in using 10 per cent. HCl. It is nothing but the ordinary dilute HCl found on every laboratory shelf. There is nothing arbitrary in using a sieve to scatter the cement upon the acid and prevent the development of sufficient heat to render a part of the silica insoluble.

Moreover, our correspondent admits that a theoretically perfect cement will contain, in spite of anything, 3 per cent. of fuel ash and 1 per cent. or less of unburned clay. (*v, w, x, z, zz, zzz*.) As these materials are not cement, why should they or any part of them be brought into solution with the cement, if it could be avoided?

Let us suppose that an unskillful or dishonest manufacturer uses more coal than is necessary or poor coal, and thus increases his coal ash to 6 per cent. and his unburned clay to 4 per cent., why should the 6 per cent. be dissolved by fine grinding or the 10 per cent. be dissolved by fusing with sodium carbonate?

In other words, why should a method of analysis be pursued that destroys all differences, while the physical tests are conducted in such a manner as to exhibit all differences?

PROOF THAT THIS CONTENTION IS CORRECT.

That this contention is correct is proved by the results of our analytical work.

Chemical Laboratory sample No. 149. Physical Laboratory sample No. 88, was a cement that gave the highest figures in physical tests. (See pages 127 and 131.)

The results of these tests were:

1 day's neat	429 pounds.
7 days' neat	784 "
7 days' mortar	209 "

When analyzed this cement gave

Insoluble in 10 per cent. HCl............   3.79 per cent.

Soluble silica	19.43	per cent.
Alumina and Iron oxide	8.34	"
Lime	63.44	"

91.21 per cent.

Magnesia	trace.	
Sulphuric Oxide SO_3	1.89	per cent.
Volatile at a Red Heat	1.39	"
Undetermined	1.72	"

5.00 per cent.

100.00 per cent.

$$\text{Active index} = \frac{63.44}{2.8 \times 19.43 + 1.1 \times 8.34} = 1.005$$

That is to say, we have here a commercial cement that consists of 91.21 per cent. cement, 3 per cent. of fuel ash, 0.79 per cent. of unburned clay, 1.89 per cent. of sulphuric oxide, 1.39 per cent. of carbonic acid and water, and 1.72 per cent. of ingredients not determined.

This is practically a perfect cement, and the other six best American cements so nearly approximated this cement in composition that with it they averaged as follows:

1 day's neat	259	pounds.
7 days' neat	683	"
7 days' mortar	257	"

Insoluble in 10 per cent. HCl .... 4 38 per cent.

Soluble Silica.	18.45	per cent.
Aluminum and Iron oxide..	9.46	"
Lime	61.89	"
		89.80 per cent.

Magnesia	1.78	per cent.
Sulphuric oxide, SO_3	1.87	"
Volatile at a Red Heat	1.79	"
Undertermined	0.38	"
		5.82 per cent.
		100.00 per cent.

$$\text{Active index} = \frac{61.89}{2.8 \times 18.45 + 1.1 \times 9.46} = 0.997$$

That is to say, these seven American Cements are 89.8 per cent. pure, with 3 per cent. of fuel ash, 1.38 per cent. of unburned clay, 1.78 per cent. of magnesia, 1.87 per cent.

of sulphuric oxide, 1.79 per cent. of carbonic acid and water, and 0.38 per cent. undetermined.

If a method of analysis can be used that will separate this unburned clay and fuel ash from the cement, why does it lead to erroneous conclusions?

Why should a method be used that makes it impossible to distinguish between the silica and alumina in the unburned clay and fuel ash, and the iron oxide that can only be dissolved in concentrated hydrochloric acid, and the remainder of those ingredients that are properly combined to form cement?

Further Tests and Proof.

In order to test these conclusions further, we wrote our correspondent and asked him to send us a sample of the coal he was using, the mix that was burned and the clinker that resulted from the burning. With great kindness, he immediately complied with our request.

We labeled the samples:

Coal	449b.
Mix	449c.
Clinker	449d.

We found the clinker to consist of small nearly spherical masses about the size of peas, that had been vitrified and that were hard under the pestle and also, larger pieces, that were more or less spongy and softer under the pestle.

449b yielded 11.95 and 11.76 per cent. of ash, an average of 11.85 per cent. which at 400 pounds to the barrel of cement, equaled 4.44 pounds to 100 pounds, or 4.44 per cent. of ash instead of 3 per cent. Of this ash 22.92 per cent. was soluble in 10 per cent. Hydrochloric Acid.

The Ash consisted of:

Silica．	42.94 per cent.
Alumina and Ferric Oxides．	41.41 "
Lime (CaO)．	9.52 "
Sulphuric Oxide SO_3	3.58 "
Undetermined．	2.55 "
	100.00 per cent.

The sulphuric oxide was wholly soluble in dilute Hydrochloric Acid.

449c contained 19.798 per cent. insoluble in dilute Hydrochloric Acid. This insoluble portion contained of Silica 68.774 per cent. and Alumina and iron oxide and lime, 31.226 per cent. These figures are equivalent to 13.62 per cent. and 6.17 per cent. of the amount taken. The soluble portion contained soluble Silica, 0.17 per cent.; Alumina and iron oxide, 0.60 per cent.; Lime (CaO), 41.16 per cent.

These results are equivalent in round numbers to:

Clay．	20 per cent.
Marl．	80 "

449d when finely pulverized gave :

Insoluble in 10 per cent. HCl	4.62 per cent. and	3.36 per cent.
Soluble Silica	18.24 " "	19.84 "
Alumina and Ferric Oxide	8.74 " "	8.65 "
Lime (CaO)	65.64 " "	65.38 "
Magnesia (MgO). Sulphuric Oxide (SO$_3$)	trace	trace
	97.24 per cent. and	97.23 per cent.
Undetermined	2.76 " "	2.77 "
	100.00 per cent. and	100.00 per cent.

These analyses were made on two separate portions pulverized separately. The insoluble portion did not contain a trace of lime.

449d. 5.0334 grams of the small unbroken clinker was placed in a flask with 250 c.c. of 10 per cent. HCl on March 15, at 3 P. M. On March 16, at 9 A. M., the action of the acid was very marked, each mass was covered with a shell of white silica. On March 18, at 9 A. M., nothing remained of most of the spheroids but a white shell of Silica. A few of them were but little acted on. On the morning of the 20th the residue of Silica was wholly white and the solution had the color of proto-chloride of iron. The contents of the flask were then analyzed and gave :

Insoluble Silica	20.95 per cent.
Soluble Silica	2.09 "
Alumina and Ferric Oxide	8.43 "
Lime	66.69 "
	98.16 per cent.

449d. 5.0908 grams of the large porous pieces of clinker were placed in a flask with 250 c.c. of 10 per cent. HCl, on March 20. The action of the acid was very unequal. Some of the pieces were completely decomposed in 48 hours; others apparently containing more iron resisted its action for several days. On March 26, one piece was still brown and was broken down with difficulty. The contents of the flask were then analyzed and gave :

Insoluble Silica	22.01 per cent.
Soluble Silica with trace of iron	1.56 "
Alumina and Ferric Oxide	8.97 "
Lime	66.72 "
	99.26 per cent.

These results show that our correspondent was mistaken in assuming that (y)

"the best Portland cements are thoroughly homoge-
" neous in character,"

as his own clinker is not homogeneous. Moreover, these analyses prove beyond question that the results of the action of acid out of the same bottle and acting on material from the same bottle depends upon the size of the masses of cement. The two samples of clinker above set forth were purposely ground of different degrees of fineness.

We then asked our correspondent to send us samples of coal, mix, clinker and ground cement, that would be as nearly as possible representative of the same batch of mix. These were received by us and numbered

        Coal .... 449 E.
        Clinker............ 449 D.
        Mix .... 449 C.
        Cement .... 449 B.

449 E yielded 10.63 per cent of ash, which is equal to 3.99 per cent. of ash in the clinker. When analyzed this ash yielded :

Silica	43.30 per cent.
Alumina and Ferric Oxide	38.69 "
Lime	9.12 "
	91.11 per cent.

There was undetermined SO_3, CO_2 and a trace of H_2S.

This analysis, as well as 449b, seems to indicate that the lime in the fuel ash combines to form a soluble compound with a part of the silica, alumina and iron, while the remainder of the silica, alumina and iron, being in excess, still continues in combination in a form not readily decomposed by dilute acids, although easily attacked by concentrated acids.

This explains why some of the very best cements are gelatinized with only a little more than a trace of matter undecomposed by concentrated acids and yield a residue to dilute acid having the appearance of ashes, but containing no lime, and less in amount than the ash contained in the coal used, as is shown in the analytical results given above.

449 D. The clinker was finely pulverized.

It contained a trace of SO_3 and no MgO.

Analyzed it gave:

Insoluble in 10 per cent. HCl	2.81	per cent.
Soluble Silica	20.61	"
Alumina and Ferric Oxide	8.69	"
Lime	65.62	"
	97.73	per cent.

449 C gave on analysis:

Silica	14.62	per cent.
Alumina and Ferric Oxide	9.21	"
Lime	41.05	"

The clay had absorbed some water. This mix is slightly different from the first, containing a little more clay.

449 B. The Physical Tests were:

1 day neat	369	pounds.
7 days neat	720	"
28 days neat	756	"
7 days mortar	209	"
28 days mortar	326	"

The cement was analyzed as usual. It gave:

	Per Cent.	Per Cent.	
Insoluble in 10 per cent. HCl	....	3.73	
Soluble Silica	19.65		Per Cent.
Alumina and Ferric Oxide	7.95		95.60
Lime (CaO)	64.27		
		91.87	
Sulphuric Oxide (SO$_3$)	1.54		
Magnesia (MgO).			
Volatile at a red heat	3.11		
		4.65	
		100.25 per cent.	

$$\text{Active Index} = \frac{64.27}{2.8 \times 19.65 + 1.1 \times 7.95} = 1.008$$

This sample when analyzed by Humphrey's method gave:

Silica	22.36	per cent.	
Alumina and Ferric Oxide	7.52	"	95.92 per cent.
Lime	66.04	"	

AN OBJECT LESSON.

449 F. A mixture of pulverized fire-brick and lime was made, having approximately the ultimate composition of a Portland cement.

When analyzed by our method the results were as follows:

Insoluble in 10 per cent. HCl............	33.25 per cent.	
Soluble Silica..........	.11 "	
Alumina a n d Ferric Oxide............	.28 "	94.53 per cent.
Lime................	60.89 "	

The insoluble portion contained:

Silica............	58.52 per cent.	19.45 per cent. of whole
Alumina and Ferric Oxide..........	29.98 "	8.97 "	"
Lime............	11.98 "	3.98 "	"
		32.40 "	"

Magnesia undetermined.

Analyzed by Humphrey's method it gave:

Silica..................	19.82 per cent.	
Alumina and Ferric Oxide	8.68 "	94.94 per cent.
Lime..................	66.44 "	

Magnesia undetermined.

Humphrey's method is not a proper method of mineral analysis:

Whatever portion of the material is soluble should be dissolved without fusion. Fusion in the manner described introduces a large amount of sodium chloride into the assay to no purpose, and should therefore be avoided. So much sodium chloride makes it difficult to separate the silica completely, and prolongs the scheme to no purpose. Moreover, the results given above speak for themselves.

Our Method of Analysis.

Believing that the results obtained by the method of analysis used by ourselves, and the interpretation of which they are capable, furnish the strongest possible argument for its general use, we submit the following suggestions respecting it.

First—That the samples should be taken and preserved in a well stoppered bottle.

Second—That the sample for analysis should be exactly like the sample submitted to physical tests. It should be subjected to no treatment whatsoever that will change its properties in any manner. It should be neither dried nor pulverized.

Third—That five grams should be weighed out and gradually introduced into 250 c.c., of not stronger than 10 per cent. HCl, in such a manner as to avoid any appreciable rise in temperature. The solution should be vigorously stirred at intervals for half an hour. Portland cements of good quality do not effervesce. If effervescence follows it indicates the presence of carbonic acid. Traces of H_2S are to be expected. A black residue indicates carbon or soot, which may be estimated by gathering the residue on a balanced filter, weighing, burning off the carbon and again weighing.

Our correspondent regards this residue as the coarse particles of the cement deprived of lime by "local separation," whatever that may mean. There is not a particle of evidence to support any such contention. The residue from 10 per cent. HCl is not coarse. It is fine as ashes, and is ashes. He says that it is not safe to add to the raw material lime enough to convert the silica and alumina of this fuel ash into cement, hence it is properly to be inferred that the lime and ash will not unite to form cement. That being the case, the cement is dissolved away from the ash and unburned clay, the former of which can only be brought into solution by prolonged action of the acid on the material in very fine powder. To us, this means which we have discovered of separating the cement from that which is not cement, seems extremely fortunate, and we cannot understand why anyone else who desires to distinguish between good and bad cements should think differently.

Fourth—That the filtered solution is evaporated to dryness over a water bath, the residue carefully and completely desiccated, drenched with concentrated HCl, warmed, taken up in water, the silica filtered off, dried and ignited as "soluble silica." This is a proper designation to be applied to this material, as it has been wholly dissolved in the action of the weak acid upon the cement. It is also the silica that is in combination to form cement, if not augmented by the silica that forms the bulk of the fuel ash, and, in Humphrey's scheme, also by the silica of the unburned clay.

Fifth—That the filtrate from the silica is made up in a graduated flask to 1 litre, and two portions of 100 c.c.

each, are precipitated with ammonium hydrate, the ammonia boiled off and the precipitate brought upon a filter and filtered while hot. Mr. Humphrey very properly dissolves this precipitate in dilute HCl and reprecipitates, passing the filtrate through the same filter. In no other way can the iron and alumina be freed from lime and magnesia, if either exists in considerable quantity.

Sixth—That the united filtrates are brought to a temperature near boiling and the lime precipitated as oxalate. After being allowed a sufficient time to settle, the lime may be filtered off and determined either as oxide or carbonate, or titrated with potassium permanganate, as suits the convenience of the operator. It is well to follow Mr. Humphrey and redissolve and reprecipitate the lime in presence of magnesia.

Seventh—That if extreme accuracy is required, the filtrate from the lime will be evaporated to dryness, ammonium salts expelled, and the magnesium precipitated with hydro-ammonium-phosphate. For practical purposes this evaporation is not necessary, but the magnesia may be determined in the filtrate from the lime.

Eighth—That for practical purposes too, the sulphuric oxide may be determined in the filtrate from the magnesia, after it has acidulated, but the original solution is better. To test this point, the sulphuric oxide was determined, (*a*) in the original hydrochloric acid solution of a cement freed from silica, (*b*) in the same solution from which the iron and alumina had been precipitated, (*c*) in the same solution from which the iron and alumina had been precipitated and the ammonia had been boiled off, (*d*) from the solution after magnesia. The determinations (*a*) and (*c*) varied by 0.03 of 1 per cent., (*d*) was $\frac{1}{10}$ of 1 per cent. greater, while (*b*) was very much less than ether of them. According to our experience barium sulphate thrown down in presence of iron and alumina always contains traces of these oxides.

Our Conclusions.

With the determination of the matter volatile at a red heat, these results furnish all of the analytical data necessary to form a judgment concerning the quality of any cement, either Portland or Rosendale, and with slight modifications the method may be applied to concretes.

To cast these results into the form of an "Active Index," or "Inactive Constituents," will not change the results in any respect, but may be useful in comparisons, as are many other wholly artificial devices.

It was originally our purpose to discuss the significance of results obtained by us in the determination of the lime and alumina of cements soluble in water, both gravimetrically and by titration, but the length of this report leads us to defer this matter till another occasion.

Respectfully submitted,

OTTO H. KLEIN,
Chief Engineer.
S. F. PECKHAM,
Chemist.

Exhibits

"A."
"B."
"C."
"D."
"E."

Exhibit "A."

Dr. Charles F. McKenna's Letter.

(*a*) " Sir—The article which appeared in your issue of the "4th inst., under the title ' Relations between Physical and "' Chemical Tests of Cement', being an abridgement of the "Report of Otto H. Klein and S. F. Peckham to the Com- "missioners of Accounts of New York City, contains so "many egregious misconceptions of the role of chemistry in "the technology of Portland cement, so many contradictory "passages and so much ignorance of cement manufacturing "processes, as to make the judicious grieve and cause an "emphatic protest against allowing such work to be ac- "cepted abroad as representative of the depth of American "knowledge of the chemistry and technology of cements. "A protest is needed for the further and more important "reason that the evils, which have resulted in this country "from a misunderstanding of the methods and their tech- "nique and purpose recommended in the physical testing "of cement, bid fair, with such a publication as this having "a vogue, to bring into the field of the chemical study of "cement a still longer train of false notions, bad methods, "encouragement of fraud, discouragement of honesty and "really serious consequences to the engineering profession."

(*b*) " The editorial comments upon that report in your "same issue pointed out a few of the absurdities of this "laborious production, as, for instance, inviting manufac- "turers and importers to furnish the 'average sample,' and "the crudeness or freshness exhibited in the treatment and "conduct of the work by the Commissioners and their "experts."

(c) "According to the authors, the object of this re-search was to ascertain 'what relation, if any, exists be-tween the results obtained by physical tests and chemical analyses of cements.' In looking for the answer to this throughout the report and in the summing up, we find 'no such correspondence was revealed in any case and the results were apparently without relation.' Nevertheless they are later and very fully discussed as if such a relation had been proven."

(d) "We find that the chemist states that he had had a large experience in cements, cement clays and limestones, but not in Portland cements, that this was supplemented by a careful examination of the last edition of 'Crooke's Select Methods of Chemical Analysis,' that the result was the practical adoption of Stillman's method and when this was abandoned his own method was adopted. It is a pity that he has overlooked such excellent articles on this branch of the subject as those contained in the treatise of Schoch and that of Candlot, but apart from this his method is genuinely unique in the annals of analytical chemistry and some of the conclusions which he has been led to in dis-cussing analyses made by it are absurd."

(e) "In the first place, he misuses the well understood term 'soluble silica,' and applies it to that portion only of silica which goes into solution in 20 per cent. hydrochloric acid, ignoring any silica of silicates insoluble in such. Now, in mineralogical chemistry, and in all analytical chemistry for that matter, 'soluble silica' is both the silica soluble in acid plus the forms of silica soluble in soda carbonate solution subsequently applied. It may be that this residue may not generally be of great value in a

" cement, but no one is justified in so asserting in any par-
" ticular case, and above all in misusing the term in a way
" that will render results useless for comparison with the
" best practice in this country and in Europe."

(*f*) " He uses 5 grams throughout the analysis, without
" subdividing, and in a cement with 60 per cent. lime he
" succeeds in determining the lime by ignition of the oxalate
" to constant weight. Most chemists would think this a
" difficult achievement since the bulk and surface of this
" precipitate will be so large as to be very favorable to the
" considerable absorption of moisture by the caustic lime."

(*g*) " It appears that in presenting his experiments the
" chemist was interested in seeing how the method of
" adding acid to the sample of cement, or vice versa,
" affected the results he obtained in percentage of lime in a
" given sample. He gives the astonishing figures 55.05,
" 62.32, 62.74, 61.76, and 61.72 per cent. all obtained on the
" same sample, and the differences to be ascribed to the
" difference in manipulation mentioned. One can under-
" stand how the amount of silica going into solution would
" differ, but why analyses of the same sample of cement
" should give these varying percentages of the total of one
" constituent is beyond my comprehension."

(*h*) " 'He says that ' the determination of the carbonic
" ' acid involved a large amount of work of questionable
" ' practicable utility, as after a number of very careful deter-
" ' minations it was found that the carbonic acid found was in
" ' every case a small percentage less than the amount vola-
" ' tile at a red heat. This latter determination was made by
" ' heating 1 gram in a small covered platinum crucible to
" ' a constant weight.' Few chemists would accede to this

" and none would of those who have had experience with
" the cements that are offered engineers to-day. And it
" does not agree with his own results, as he mentions No. 30
" giving 12 per cent. carbonic acid, while his analytical table
" (not included wholly in your text, but issued in a blue
" print at the meeting of the Society of Chemical Industry
" at which the paper was presented) shows that the amount
" volatile at a red heat on this sample was 17.92 per cent.
" No. 58 gave 28.86 per cent. volatile matter and he does not
" class it with those adulterated with carbonate of lime, dis-
" cussed elsewhere in his paper, but says of it: No. 58 is a
" poor cement all through, being low in lime, high in mag-
" nesia, with nearly one per cent. of carbon, and showing by
" the loss at the red heat of 28.86 per cent., either the effects
" of ageing or bad or insufficient burning. It is a very bad
" cement."

(i) " The carbon he finds he ascribes in all cases to un-
" burned coal, even though it is quite likely that many of the
" cements which he examined had been burned in rotary
" kilns using oil fuel. He does not seem to know that at
" times some manufacturers purposely grind a small portion
" of coal with the clinker."

(k) " His determination of lime and alumina soluble in
" water was rightly called important, probably more impor-
" tant than he seems to think, but he does not mention the
" valuable and simpler standard alkali metric test of a water
" solution of 0.5 gram of cement with decinormal acid.
" In fact, how unique this experimenter is in his chemical
" work may be infinitely clear to engineers familiar with
" physical testing when they read how he adds together all
" the figures for tensile strength at different periods and uses
" the sum as a means of classification."

(*l*) "As to his use and misuse of the term 'active index,'
"I will only say this: The active index is only one of many
"interrelations of composition and character which are used
"for diagnosis and which are to be credited with more or
"less weight. It is a figure which should be delicately
"touched upon, which has been derived from more or less
"uncertain assumptions of the presence of the bases and
"acids in certain definite combinations and which assuredly
"cannot be adopted for judgment of a cement which has
"been analyzed according to such a method as we find here.
"At least, the adulterants should be ascertained to exist in
"a certain form in a cement before gray matter is wasted
"over 'active index.'"

(*m*) "But what shall be said of the attempt here made to
"apply the theoretical formula for pure Portland cements to
"natural cements. This is so shockingly absurd, so demon-
"strative of an ignorance of the very nature of hydraulic
"action and of the character of the various hydraulic mate-
"rials, that nothing need be said further than to quote him
"where, applying Newberry's formula to Rosendale cements,
"he says: 'In general also they exhibited the effects of
"'both under and over burning.'"

(*n*) "Apologizing for the taking of so much space in your
"columns, I will close with the statement that there is much
"other similarly interesting material in this report and that
"I have tried to set down here nothing that would be unfair
"to the authors of it, while yet justifying my assertion that
"its publication is a wrong to American chemical and engi-
"neering practice."

"Very truly yours,
"(Signed)   CHAS. F. MCKENNA."

Exhibit "B."

EDITORIAL CRITICISM OF *Engineering Record*

QUOTED BY MR. LEWIS.

"As a matter of fact the main criticisms which may be
"made on the report can be based on a certain kind of crude-
"ness or freshness exhibited in the treatment and conduct
"of the work by the Commissioners of Accounts and their
"experts. It may even be doubted whether all the positions
"taken by them can be successfully defended. They have
"made investigations of value to The City of New York,
"and their efforts and purposes are to be commended, but
"some of their operations should have been conducted in a
"different manner * * * Again, the rather sweeping
"rejection of the prevailing methods of chemical analysis of
"cements in this country and of Europe is at least open to
"some question."

Exhibit "C."

QUOTATIONS FROM
ELEVEN PAGE ANSWERING REPORT,
DATED JANUARY 4, 1901,
MADE BY
ENGINEER OF THE DEPARTMENT OF HIGHWAYS,
BOROUGH OF BROOKLYN,
ON
WATKINS STREET REGULATING AND PAVING CONTRACT.

After showing that the physical tests applied to the cement in question gave results that were quite variable, but averaging above the required strength for Rosendale cements, he proceeds:

1. "Chemical analyses have been made of the "samples taken on October 23, and on November 5, in "order to compare them with those obtained by "Professor Peckham, and given in his report marked "Exhibit "A."

"These analyses showed the following composition:

	SAMPLE OF OCTOBER 23.	SAMPLE OF NOVEMBER 25.
Carbon...	.985	1.022
Matter insoluble in 10 per cent. HCl...	9.486	13.277
Magnesia...	2.843	2.861
Sulphuric Oxide...	.760	.759
Matter volatile at Red Heat...	13.369	14.972
Inert constituents...	27.443	32.891
Soluble silica...	14.741	13.128
Alumina and Iron Oxides...	5.728	5.439
Calcium oxide (lime)...	52.102	48.520
	100.014	99.978

2. " It will be noticed that the most conspicuous
" difference between the above analysis and that of
" Professor Peckham is in the larger percentage of
" matter insoluble in 10 per cent. hydrochloric acid, and
" in the materially larger per cent. of lime in our two
" samples.

3. " The report received from our Chemical Labora-
" tory with these analyses, states as follows :

" As will be seen above, the percentages of silica,
" alumina and iron oxides and lime (which are the
" active hydraulic ingredients of a cement) are obtained
" by decomposing the silicates, aluminates, etc., of the
" cement with 10 per cent. *Dilute* Hydrochloric acid.
" Admitting for the sake of argument that the percent-
" ages of silica, alumina and ferric oxides, and lime, so
" obtained represent the maximum percentages of active
" hydraulic ingredients, the only conclusion that we can
" draw is, that these analyses show a cement in which
" the active hydraulic ingredients are admirably well
" balanced. The percentage of inert constituents is
" very low, as shown by comparison with the average
" Rosendale Cement analysis in the report of The
" Commissioners of Accounts of May 24, 1900. The
" averages of the percentages of inert constituents of
" Rosendale Cements is 40.22 per cent."

4. " We however believe that the above method of
" analysis to be a new, radical and arbitrary one, and at
" variance with the usual practice, which is to employ
" *concentrated* hydrochloric acid, and determining all of
" the silica, alumina, and iron oxides, and lime so

" decomposed from the silicates, aluminates, etc. This
" method has given entire satisfaction in the past, and
" we see no reason for discarding it."

5. " In using concentrated hydrochloric acid, the un-
" decomposed silicates, etc., are reduced from 9.5 to 2.3
" per cent. and from 13.3 to 3.8 per cent., while the solu-
" ble silica is increased from 14.7 to 19.6 per cent., and
" from 13.1 to 20 per cent., and the lime is increased a
" little over 1 per cent. The report of our chemist, Mr.
" Broadhurst, then proceeds as follows:

" It has never been claimed, that Rosendale (nat-
" ural) cements contain ingredients in such proportions
" as to make a theoretically perfect cement. While
" chemical analysis is an invaluable aid in determining
" deleterious matter and the cause of disintegration of
" cements, there are certain combinations due entirely to
" the method of burning, which render a cement valuable
" or worthless, as shown by physical tests and which
" chemical analysis fails to indicate."

6. " We consider this Commercial Rosendale Ce-
" ment to be a very superior one for the following rea-
" sons:

"Comparatively small percentage of inert constituents;

" Low percentage of magnesia;
" Low percentage of sulphuric Oxide;
" Soundness as shown by the boiling test;
" Great ultimate tensile strength."

7. " The Commissioners of Accounts claim that the
" conclusions drawn from tests and analyses made of the
" sample of cement were confirmed by chemical
" analyses, made of some of the concrete taken from the
" street on November 28, 1900, by Mr. Creuzbaur,
" representing the Department of Finance, and Messrs.
" Klein and Dusenberry, representing the Commission-
" ers of Accounts."

8. " The manner in which these so-called samples
" were taken, the analyses that were made of the mortar
" and the comparison of the results obtained with a
" purely hypothetical standard, are such as to render
" their conclusions entirely valueless. The concrete on
" the street in two small spaces, each one foot or less in
" diameter was broken up. The loose material con-
" sisting of stone and mortar, was scooped out. The
" material from each of the holes was then divided into
" two parts, one part of each being taken to the Chemist
" of The Commissioners of Accounts for Analysis. In
" the report (Exhibit " B ") it is stated that they ' were
" ' brought in boxes ' and ' the lumps of stone were
" ' loose in the sand and cement.' A portion of the
" material was poured from each box and the
" stone was separated from the fine material. In
" the case of one portion examined the stone
" weighed 360 grams, and in the other case 547
" grams, and it would appear as though the chemist
" had first intended to examine the concrete as a whole,
" but he subsequently confined his investigations to
" the mortar only. This mortar was treated with dilute
" hydrochloric acid, and from the result of the analyses
" it is found that one sample contained 22 per cent. and

" another 19.5 per cent. of active matter. These results
" are then compared with the amount of active matter
" which should be found in a mortar made of one part
" of cement, containing 1.5 per cent. of inert matter, and
" two parts of sand. Their attention is called to the
" fact that the analysis of the Standard Rosendale
" cements *furnished by the manufacturers* as de-
" scribed in the report of the Commissioners of
" Accounts of May 24, 1900, contained about 46 per
" cent. of inert matter, and the ordinary run of Com-
" mercial Rosendale cement contains about 30 per cent.
" of inert matter, the absurdity of comparing the re-
" sults obtained from an analysis of the mortar found
" on Watkins Street with a hypothetical Rosendale
" cement containing only 15 per cent. of inert matter
" is so apparent as to need no comment."

9. " In order to make a fair examination of the con-
" crete, the asphalt pavement was opened in two places
" on December 29, 1900, one opening about two feet
" square being made in about the middle of the block
" between Sutter and Blake avenues, and one about
" four feet square in the block between Blake and
" Dumont avenues. These openings were made in the
" presence of Mr. R. W. Creuzbaur, Principal Assistant
" Engineer of the Department of Finance, Mr. George
" W. Tillson, Principal Assistant Engineer of this
" Department and myself. Upon removing the asphalt,
" the concrete was in both cases found to be in good
" condition, firm, dense, and while not, of course, as
" hard as would have been a concrete made of Portland
" cement, it was in the condition to have been expected
" of concrete made of good natural cement. It resisted

" the blows of a pick until the mass was shattered, when
" it could readily be broken. In the second opening a
" large unbroken slab of concrete was cut out. Two
" portions of this were brought to this office, one of
" them for Mr. Creuzbaur and the other in order that
" a fair separation of the stone and mortar could be
" made from a piece large enough to furnish some indi-
" cation as to the manner in which the concrete was
" mixed. The results of this examination conducted
" by Mr. W. H. Broadhurst, Chemist of this Depart-
" ment, along the lines described in the report of
" Professor Peckham, known as Exhibit ' B,' were as
" follows :

" Sample H59 weighed 1,059.09 grams, of which
" 668.50 grams were stone and 390.59 were mortar.
" Sample H60 weighed 3,794.05 grams, of which 2,563.95
" grams were stone and 1,230.10 grams were mortar."

10. " Mr. Broadhurst's report continues as follows :

	Sample H59.	Sample H60.
Insoluble Matter.		
Sand and other mineral matter.	71.180 per cent.	71.342 per cent.
* "So-called" carbon	.703 "	.831 "
	71.883 per cent.	72.173 per cent.
Soluble Matter.		
Silica	3.620 per cent.	3.461 per cent.
Alumina and Iron Oxides	2.846 "	2.810 "
Lime	13.780 "	13.462 "
Magnesia	Trace	Trace.
Sulphuric Ox'de	Trace.	Trace.
	92.129 per cent.	91.906 per cent.

* " Insoluble matter volatile on ignition."

"We find only a trace of iron insoluble in con-
"centrated hydrochloric acid, there is hence none
"rendered inert by burning."

11. "A mortar made of one part of Rosendale ce-
"ment and two parts of sand by volume is equivalent
"to 28.67 parts of cement to 71.33 parts of sand by
"weight (computed from results given in Annual Re-
"port of Chief of Engineers, U. S. A., 1894, page
"2313). 1 cubic yard of mortar—2.75 bbls. of cement
"of 300 lbs. per bbl. (80 lbs. per cubic foot). .76 cubic
"yards of sand, 100 lbs. per cubic foot."

12. "This Commercial Rosendale cement contains
"approximately 30 per cent. of so-called 'Inert Mat-
"ter' (Matter insoluble in 10 per cent. Dilute Hydro-
"chloric acid). We do not know of any Rosendale
"cement which contains 15 per cent. of matter cor-
"responding to this figure."

"By weight, therefore, 100 parts of mortar would
"contain 71.33 parts of sand and 8.60 parts of so-called
"inactive cement, equivalent to 79.93 parts of inert
"matter and 20.07 parts of active matter."

"The above analyses show that sample H59 con-
"tains 20.246 parts of active matter, and sample H60
"19.733 parts of active matter, in 100 parts of mortar,
"corresponding almost exactly with the proportions of
"sand and cement required by the specifications."

13. "The results which I have given you of tests
"which have been carefully made demonstrate, I think,
"beyond a doubt that the concrete on Watkins Street

" was laid in accordance with the specifications, and the
" results are satisfactory. You will note that the con-
" clusions drawn by our chemist are based largely upon
" premises laid down by the Commissioners of Ac-
" counts themselves, although it is not admitted that
" these premises are sound. The whole theory of their
" investigations of Rosendale cements being carried
" along the same lines as are investigations of Portland
" cements is certainly untenable. To show that this
" Department is not alone in this opinion, I beg to
" refer to the editorial comment on the much-advertised
" report of the Commissioners of Accounts of May 24,
" 1900, which is contained in the *Engineering Record*
" of August 4, and which says :

Exhibit " D."

QUOTED AND DISCUSSED AS EXHIBIT " A " IN THE LETTER OF MR. LEWIS.

NEW YORK, N. Y., December 7, 1900.

OTTO H. KLEIN, Esq., *Chief Engineer,*
  Commissioners of Accounts, New York:

MY DEAR SIR—Concerning the sample of cement marked No. 279, I have to report as follows:

14. The reaction indicated a cement that was badly burned. The finest ground portion that went through the sieve first did not effervesce; but the coarser particles that went through the sieve last and some coarse fragments that did not go through a No. 40 sieve at all, effervesced like fresh limestone. The portion insoluble in 10 per cent. hydrochloric acid contained a considerable amount of grains of quartz sand and insoluble oxide of iron, indicating that some portion of the cement was overburned and that there was not the necessary care taken in the manufacture to exclude the inert sand that helped swell the proportion of the inactive ingredients to very near one-third the whole amount of the cement.

Of these inert constituents there was:

Carbon	0.182	per cent.
Matter insoluble in 10 per cent. HCl	16.738	"
Magnesia	2.300	"
Matter volatile at a red heat	13.170	"
	32.390	per cent.

Of the active ingredients, there were:

Soluble Silica	15.065	per cent.
Alumina and Iron Oxides	6.970	"
Calcium oxide (lime)	45.540	"
	32.390	"
	99.965	per cent.

15. At least 12 per cent. of the inert matter volatile at a red heat is carbonic acid, which in the cement is in combination with a part of the lime as wholly inert carbonate of lime, amounting to 27.276 per cent. Deducting the 15.276 per cent. of lime in this inert combination from the 45.54 per cent. of lime found, there remains only 30.264 per cent. of active lime, which is just about one-half the amount required for a good cement. When such a cement as this is mixed with two parts sand, the one-half part of cement that it really contains is mixed with four half parts of inert matter in addition to the one-half part that is already there, making five half parts to one-half part or a proportion of one to five instead of one to two. It is no wonder that a concrete made from such cement is worthless.

16. The active index of this cement is 0.913, showing too little lime if it was all active. The actual amount of active lime present would show a still smaller fraction.

The following is an exhibit of the results of the analysis :

Carbon	0.182	per cent.
Insoluble matter	16.738	"
Soluble silica	15.065	"
Alumina and Iron oxides	6.970	"
Calcium Oxide (lime)	45.540	"
Magnesia	2.300	"
Sulphuric Oxide	trace.	
Matter volatile at a red heat	13.170	"
	99.965	per cent.

Active Index	0.913
Inactive Index	32.39

Respectfully submitted,

(Signed)   S. F. PECKHAM,
Chemist.

Exhibit "E."

QUOTED AND DISCUSSED AS EXHIBIT "B" IN THE LETTER OF MR. LEWIS.

NEW YORK, N. Y., December 3, 1900.

OTTO H. KLEIN, Esq., *Chief Engineer,*

Commissioners of Accounts, New York City:

MY DEAR SIR—Concerning the two samples of Concrete taken from Watkins Street marked No. 280A and No. 280B, I have to report as follows:

17. The two samples were brought in boxes. The lumps of stone were loose in the sand and cement. A portion was poured from the boxes and air dried. When dried they were sifted through a No. 10 sieve. Nearly all of the cement and sand was in this way separated from the stone. The few lumps that were coherent were easily broken up between the fingers and a complete separation made from all but a very few small fragments of stone that passed the sieve.

18. When air dry, the stone in No. 280A weighed 360 grams and the sand and cement 235 grams. In round numbers 60 per cent. and 40 per cent. respectively.

When air dry, the stone in No. 280B weighed 547 grams and the sand and cement 285 grams. In round numbers 65 per cent. and 35 per cent. respectively.

19. The two samples were analyzed air dry, as if they had been cement, the material separated from the stone being in the form of a coarse powder.

Ten grams of each sample were taken. When introduced into dilute hydrochloric acid both samples effervesced freely from escape of carbonic acid gas. The matter insoluble in the acid consisted of:

	280A.	280B.
Sand and other mineral matter.	69.604 per cent.	71.353 per cent.
Carbon.	.521 "	.632 "
Total.	70.125 per cent.	71.985 per cent.
The soluble matter consisted of:		
Soluble silica.	3.685 per cent.	3.385 per cent.
Alumina and Iron Oxides.	5.855 "	3.565 "
Lime.	12.440 "	12.590 "
Magnesia.	Trace.	Trace.
Sulphuric oxide.	"	"
	92.105 per cent.	91.525 per cent.

There was a large percentage of iron in the insoluble residue that was rendered inert by burning.

20. A concrete made of one part Rosendale cement, containing 15 per cent. of inert matter, and two parts sand, would contain 215 parts of inert matter in 300 or about 71 per cent. The remaining 29 per cent. should consist of cement; that is of soluble silica, alumina and iron and lime, in proper proportions. In these samples there is 22 per cent. in No. 280A and 19.5 per cent. in No. 280B. There is

not lime enough to form a hydraulic compound with the soluble silica and alumina and iron in either, provided the cement was thoroughly burned and all of the lime chemically active; but, a large part of the lime is unburned, as indicated by the escape of a large quantity of carbonic acid gas, and consequently a large part of the lime is inert and not in an active form.

In other words, there is not cement enough in the concrete to hold the concrete together, hence the unsatisfactory condition of the concrete now, in consequence of a lack of and a condition of ingredients that no amount of ageing can remedy.

Respectfully submitted,

(Signed)   S. F. PECKHAM,
Chemist.

COMPARISON BETWEEN

PHYSICAL TESTS AND ·CHEMICAL ANALYSES

OF 34 SAMPLES

OF PORTLAND AND ROSENDALE CEMENTS,

WITH THREE TABLES OF RESULTS.

REPORT

TO THE

Hon. ROBERT A. VAN WYCK, Mayor,

MADE BY

JOHN C. HERTLE,
EDWARD OWEN,
Commissioners of Accounts of the City of New York,

MAY 24, 1900,

And read before the New York Section of the Society of Chemical Industry of England, May 25, 1900.

LETTER OF TRANSMISSION.

Office of the Commissioners of Accounts,
Stewart Building, No. 280 Broadway,
New York, May 24, 1900.

·Subject—Comparison Between Physical Tests and Chemical Analyses of Portland and Rosendale Cements.

Hon. Robert A. Van Wyck, *Mayor :*

Dear Sir—" The Commissioners of Accounts of the City of New York" beg to submit herewith for your consideration a report dated May 21, 1900, made to them by Mr. Otto H. Klein, their Chief Engineer, and Professor S. F. Peckham, Chemist in charge of their Laboratory.

This report is the product of much labor on the part of these two officials, frequently extending beyond their usual hours during the period between the 26th day of May, 1899, and the present date, and embodies the physical tests and results of chemical analyses made of the following samples of cement :

	Samples.	
Domestic Portland, obtained from manufacturers..	17	
Imported Portland, obtained from agents.........	2	
		19
Rosendale, obtained from manufacturers..........	13	
Rosendale, obtained from City Works in course of construction...............................	2	
		15
Total number of samples........		34

These samples were obtained by us in reply to letters sent by this office, one of which we copy as follows :

"NEW YORK, May 26, 1899.

"GENTLEMEN—Will you be kind enough to send us, within the next "week, a twenty-five pound sample of your fresh-ground Portland "cement, the same to be used for a comparative test as a record in this "Department.

"Respectfully,
(Signed), "JOHN C. HERTLE,
"EDWARD OWEN,
"*Commissioners of Accounts.*"

In the course of our examinations of the various contracts with the City, for regulating, grading and paving, during the years 1898 and 1899, your Honor will no doubt recollect the numerous reports made by us of the poor quality of cement used in concrete foundations for roadways, which finally culminated in our recommending, in a report dated May 4, 1899, the exclusive use of Portland cements, from which report we quote as follows, viz. :

"NEW YORK, May 4, 1899.

"*Hon.* ROBERT A. VAN WYCK, *Mayor :*

"DEAR SIR—We beg to call your attention to the accompanying "cement test sheet, which shows that the Rosendale cement used in the "concrete foundation for asphalt blocks does not come up to the require-"ments of a good, serviceable cement. * * *

"The concrete foundations for pavements made of Rosendale cements "in this city have, of late, given reasons for numerous complaints, criti-"cisms and legal actions.

"We believe it would be in the interest of the Department of High-"ways, and of the Corporation Counsel, if the use of Rosendale or nat-"ural cements for foundations of pavements in this city would be "excluded from the specifications in the future, and that only the freshly "ground American Portland cements be permitted to be used.

"The concrete made thereof should be composed of one part of Port-"land cement to three parts of sand to seven parts of broken stone, the "stone to be of such a size as to pass through a two and one-half inch "ring in any direction.

"There is very little difference in the cost of concrete foundations, "whether made of Rosendale or of Portland cements, but too much "stress cannot be laid on the advantage derived from the use of Portland "cement concrete for the foundation of pavements.

"Respectfully submitted,
(Signed), "JOHN C. HERTLE,
"EDWARD OWEN,
"*Commissioners of Accounts.*"

On October 16, 1899, we made an additional report, recommending the substitution of Portland for Rosendale cement, and in said report called attention to the fact that our conclusion was also shared by the Engineers of the Comptroller's office, as is shown by the following quotation, viz.:

"NEW YORK, October 16, 1899.

"*Hon.* ROBERT A. VAN WYCK, *Mayor:*

"DEAR SIR—The former specifications made up by the Department of
"Highways for asphalt pavements on concrete foundations, for all the
"boroughs, permit the employment of both Portland and Rosendale
"cements in the concrete foundation.

"On account of the inferiority of Rosendale concretes, this Bureau, as
"well as the Engineering Bureau of the Comptroller's office, has had
"occasion to make numerous complaints, and the final conclusion was
"that the practice of using Rosendale cement in concrete foundations
"for asphalt pavements should be entirely discarded, and that the clause
"permitting the use of Rosendale cement should be eliminated from the
"specifications.

"To our great satisfaction, our recommendation to permit the use of
"Portland cement exclusively was adopted in the new specifications,
"published in May, 1899, and in our opinion, consequently, the continu-
"ous complaints and criticisms made by us in regard to bad concrete
"foundations appeared to and promised to be a thing of the past. * * "

Copies of said reports of May 4 and October 16, 1899, were, by your Honor, transmitted to the Department of Highways, and shortly thereafter we were visited by several of the cement manufacturers, and the result of these interviews prompted us to send for these samples, and make a careful study of the subject of cements.

Attached to this report will be found Tables Nos. 1 and 2, showing the results we obtained from *physical tests*, and also Table No. 3, showing the results from the *chemical analyses;* of the thirty-four samples of cement which are the subject of the following report (See pages 127, 129 and 131).

In these tabulated statements the samples are designated by a separate series of numbers for each test, for the purpose of not disclosing their identity.

To assure your Honor of the accuracy of the attached report and tabulated statements, and to show our belief in their scientific importance, we beg to state that we have consented to the reading of the report, by the authors, before the *New York Section* of the Society of Chemical Industry of England.

The usual custom of this society being to have all important papers that are read before it published in its official Journal (which has a large circulation throughout the world) we will, upon receipt of same, transmit a copy to your Honor.

We realize the fact that up to the present time, so far as we have been able to discover, no *correspondence* has been observed between *physical tests* and *chemical analyses* of cements.

This lack of correspondence appeared to us, after the thirty-four samples had been analyzed, and, as a consequence, we made the *method* of analysis a subject of investigation, and developed a new process of analysis which, upon being applied to these thirty-four samples, the *physical tests* and *chemical analyses* showed corresponding results as to the quality of each sample of cement.

The report, which we herewith submit, will, we believe, show beyond a doubt, and demonstrate our finding, that the *chemical analyses* of cements will always confirm the *physical tests*.

The report also demonstrates the correctness of our previous contentions that, in view of the price at which Portland cements can now be bought, Rosendale cements, on account of their lack of uniformity, should be entirely eliminated from the specifications in the construction of concrete foundations for roadways, where permanence and solidity are the first considerations.

The importance to this City of this result may be best understood by the statement we now make, namely, that if called upon by the Corporation Counsel, we are prepared to demonstrate the truth of our finding in a court of law.

We are at present engaged in a still further investigation, the results of which will be embodied in a separate report, of these same thirty-four samples, to ascertain if *optical examinations* will confirm the finding herein shown of both the *physical tests* and *chemical analyses*.

Respectfully submitted,

JOHN C. HERTLE,
EDWARD OWEN,
Commissioners of Accounts.

REPORT OF

MESSRS. OTTO H. KLEIN and S. F. PECKHAM.

ENGINEERING BUREAU,
OFFICE OF THE COMMISSIONER OF ACCOUNTS,
NEW YORK, May 21, 1900.

SUBJECT—Cement Testing.

Hon. JOHN C. HERTLE and EDWARD OWEN, *Commissioners of Accounts:*

GENTLEMEN—The employment of cements in the proper construction of foundations for road beds, prompted the Department of the Commissioners of Accounts of the City of New York to undertake a general examination of the most well known cements on this Market.

For this purpose several of the most prominent manufacturers of cements in the United States, as well as the importers of several foreign brands, were asked to furnish this office with an average sample of their commercial product. Thirty-four samples of thirty brands were obtained in this way, of which seventeen were Portland cements and thirteen were Rosendale or Natural cements.

All of these samples were tested in the Physical and Chemical Laboratories in series of tests, each series being carried through by the same person, in order to eliminate as far as possible the variations due to personal equation.

The results thus obtained have been tabulated and are shown on the three tables hereto annexed.

In the Physical Laboratory they were tested according to the methods recommended by the Committee on Cement Examinations of the American Society of Civil Engineers, (Transactions Am. Soc. C. E., Vols. XIII. and XIV.) which includes the determination of (1) Average Tensile Strength in pounds per square inch; (2) Fineness; (3) Activity determined by Normal Needle, and (4) Faija's Test of Soundness.

(1) Average Tensile Strength.

The instrument used was Fairbanks' Automatic Cement Testing Machine.

The determinations were conducted as follows: Three different tests were made of each sample, two neat—without admixture of any foreign materials—at periods of one and seven days, and one with admixture of sand, for a period of seven days. The proportion of cement to sand in the mortar briquettes were for Portland cement, one part of cement to three parts of sand and for the natural cement one part of cement to two parts of sand. The sand used was crushed quartz, commercially known as "No. 3." The method of mixing pursued was as follows: The amount of cement intended for any batch was carefully weighed, placed on a glass plate and mixed with sufficient water so that after having been properly compressed in the mould to remove air bubbles, the water could be made to flush the surface. By using the trowel properly, the moulds were then turned over and the same operation proceeded with on the reverse side. The weighing of the cement was by the metric system and the water used in mixing (clean hydrant water) was measured in a glass cylinder graduated to cubic centimeters, the water being kept as near as possible at the temperature of the air. While the briquettes were allowed to remain in the air they were covered with damp blotters. All of the briquettes, except those used for the one day tests, were allowed to remain in the air twenty-four hours. The one day briquettes were allowed to remain in the air a certain period, depending upon their activity. For the natural cements this period was one hour and for the Portland cements three hours. If they were not thoroughly set in that time they were allowed to remain longer in the air. There was one extreme case (No. 68 of Table No. 1) that set so slowly that it was impossible to put the one day briquette in water without its disintegrating so as to make a test impossible. Consequently, the results obtained for this cement, for twenty-four hours, were obtained from briquettes that were not immersed. In the table showing Average Tensile Strength, the averages for all briquettes were made from testing six briquettes. The percentage of the water used in mixing and the temperature of the same, are the averages of the quantities used for the one and seven day neat briquettes. The percentages used for mortar briquettes were computed on the basis of the entire weight of the mixture of sand and cement.

All of the briquettes were tested on the same Fairbank's Automatic Machine, of the latest type, of a capacity of 800 pounds. The pressure was uniformly applied at the rate of 425 pounds per minute, as determined by several experiments.

Further details could be given, but we deem this information sufficient for the explanation of the tables.

(2) Fineness.

To determine the fineness of the samples the percentages passing the standard sieves, as described below, were ascertained. One hundred grams of cement were used for each determination. The standard sieves above referred to were as follows: No. 50 sieve has twenty-five hundred meshes to the square inch and is made of No. 35 Stubb's wire gauge. No. 74 sieve has five thousand four hundred and seventy-six meshes to the square inch and is made of No. 37 wire. No. 100 sieve has ten thousand meshes to the square inch and is made of No. 40 wire.

The Portland Cements were found to be invariably more finely ground than the Natural Cements, as will be seen at a glance from the table. All the average percentages are considerably above the usual requirements of fineness. While it cannot be denied that fineness is an important factor, an inspection of the table shows that the Portland cements are uniformly fine without regard to quality. This quality of cements is discussed further on in our paper.

(3) ACTIVITY.

The activity, or time required for setting, was determined by the use of the Vicat apparatus, which may be described as follows: A right section of a hollow cone of hard rubber, of about seven centimeters average clear diameter and four centimeters in height, set on a glass plate, holds the cement paste. This cone is placed under a frame which supports in a position at right angles with the surface of the cement a movable rod, to the lower end of which may be attached a piston or needle. The piston has a diameter of one centimeter and the needle a cross-section of one square millimeter. Suitable weights are attached to the upper part of the rod, so that no matter whether either the piston or needle may be in use, the weight of the whole movable portion shall be three hundred grams. There is always attached to the rod an index moving over a scale graduated to millimeters and fiftieths of an inch, the scale being fixed on the framework.

All of the cements tested with this apparatus were brought to a uniform consistency by mixing with a certain quantity of water that the piston, as described, would sink to the base through the paste to within from six to ten millimeters of the bottom of the rubber cone, as denoted by the index.

The "initial set" was the time that elapsed between the mixing and the first moment that the needle did not entirely penetrate the paste. The "hard" or "final" set was the time that elapsed between the mixing and the moment when the paste would support the needle without an appreciable impression, which latter, in the case of some of the slow setting cements, was sometimes difficult to determine. The test above described is largely a comparative one, and is used to determine which cement will best fulfill the conditions under which it may be used. The time of initial set is also of importance as representing the limit of time within which the cement must be handled or worked. The "hydraulic activity" is the time that elapsed between the initial and final set. The results of this test are shown in the table.

(4) Faija's Test.

The test for soundness or constancy of volume consists of exposing small "pats" of cement, about three inches in diameter and of a thickness of about one-half an inch at the centre, tapered to very thin edges, and mounted on a glass plate, to a moist heat of from 100 degrees to 105 degrees F., for six or more hours, until thoroughly set; they are then immersed in a water bath for the remainder of twenty-four hours, said bath being kept at a temperature of from 115 degrees to 120 degrees F., by means of a thermostat. The purpose of this test is to impart an artificial age to the cement, during the progress of which certain defects are made apparent. If the pat adheres to the glass plate and does not crack or show any blow-holes the test may be considered as satisfactory.

A new apparatus, constructed by Professor Bauschinger, for the determination and measurement of the change of volume of cements while setting, has been added to the Physical Laboratory of the Commissioners of Accounts, but it was not received in time to enable us to apply it to all the cements in this list. We propose to give these results at a later date.

We have encountered in our physical tests results which have led us to the study of hardened cements in thin sections under the microscope. In consequence of the tedious detail of this work and also of unexpected delays over which we have had no control, we have been unable to present the results of this research at this meeting as we expected. Without doubt we shall be able to present them with the results of other experiments now in progress at an early date next year.

We are confident that a combination of physical, chemical and optical tests will ultimately reveal the cause of failure in the application of cements, hitherto regarded as obscure and difficult of explanation.

CHEMICAL ANALYSIS.

The object of this research was to ascertain how far, if at all, the results of the chemical analysis of cement supported or confirmed the results obtained in the physical examination of cements; that is to say, the question to be answered was, What relation, if any, exists between the results obtained by physical tests and chemical analysis of cements?

On looking up the literature relating to the analysis of cements it was found to be in a very unsatisfactory condition. There was found to be no agreement as to the purpose of such analyses. In general it may be said that the manner of the analyses described indicates that it is desirable to set forth the content as consisting of the largest possible percentage of lime, silica and alumina, the latter usually including the iron as ferric oxide. Such analyses are wholly in the interest of the manufacturers of the cement.

One of us has had a large experience, several years ago, in the analysis of cement and cement clays and limestones. None of these materials, however, were used in Portland Cements; they were, strickly speaking, hydraulic lines; yet the problems presented by Portland Cement are practically the same, provided an ultimate analysis is sought. The same problems were encountered in the analysis of the Estherville meteorite, which, at the time it fell, led to a long correspondence with Dr. J. Lawrence Smith, which his wide experience made extremely valuable. All of this experience was supplemented by a careful examination of the methods described in the last edition of Crooke's Select Methods of Chemical Analysis. The result was the practical adoption of Stillman's Method of Cement Analysis, as described in the Journal of the American Chemical Society for April, 1893, and March, 1894.

This scheme is practically as follows: Drying the material and weighing out two grammes; solution in 50 c.c. CHl and 5 c.c. HNO_3 (presumably concentrated); evaporation to dryness; addition of 25 c.c. HCl. (conc.?) and 100 c.c. of H_2O, and boiling; filtering into ¼ litre flask and making up solution to ¼ litre; analyzing residue by fusion and determining silica and alumina; making 100 c.c. of solution alkaline, with NH_4HO, warming, filtering and washing, drying, igniting and weighing precipitate as $Al_2O_3 + Fe_2O_3$; fusing this precipitate into a silver dish, with KHO, weighing the Fe_2O_3 and determining Al_2O_3 by difference; adding to filtrate $(NH_4)_2C_2O_4$ in slight excess, setting aside for four hours, filtering, washing and determining lime after ignition over a blast lamp as CaO. The remainder of the scheme is devoted to an unnecessarily intricate determination of magnesia, sulphuric oxide and the alkalies.

Some preliminary work led to the planning of a schedule that should set forth: the laboratory number, total calcium oxide, calcium oxide soluble in water, silica soluble and insoluble, aluminum and iron oxides, aluminum oxide soluble in water, iron as ferric oxide, metallic iron from grinding machinery, magnesium oxide, sulphuric oxide which is equivalent to sulphate of lime, carbonic acid which is equivalent to carbonate of lime, loss at a red heat, carbon and water.

From this scheme the alkalies were soon eliminated as involving a large amount of labor, practically to no purpose. We soon determined that the insoluble portion was not silica ; also that concentrated acid was not fit to be used, as it rendered the silica insoluble; also that iron from the grinding machinery, while present in the sample that we first took up, was absent from nearly all the others. It was, therefore, eliminated from the scheme. We thereafter soon concluded that an ultimate analysis did not reveal differences in very unlike cements. Further, on a careful review we found that the use of a 20 per cent. solution of HCl was generally recommended without specifying the manner of its use.

We then proceeded to weigh out five grammes of the specimen without drying, as drying often greatly changes the specimen from the material actually submitted. This was placed in a casserole and about 100 c.c. of a 20 per cent. solution of HCl was poured upon it. The reaction varied, but was usually very vigorous, with evolution of heat sufficient to raise the temperature to about 100° C. The insoluble residue was filtered off, washed in hot water, dried and ignited as "insoluble matter." The solution was then evaporated to dryness over a water-bath, the residue dried at 120° C., cooled, drenched with concentrated HCl, allowed to stand half an hour, diluted with about 200 c.c. of water, filtered, and the residue washed with hot water, ignited and weighed as "soluble silica." The solution was then diluted to one litre and equal portions of 100 c.c. were rendered alkaline with NH_4HO, diluted with water and the ammonia boiled off. The precipitate, consisting of Al_2O_3 and Fe_2O_3 with traces of silica, lime and magnesia, was filtered off, dissolved on the filter with diluted HCl, the filter well washed with hot water, the acid filtrate carefully precipitated with NH_4HO, filtered through the same filter, washed, ignited and weighed. A trace of silica follows the precipitate, and can only be removed by putting the dilute acid solution through the filter a second time and gathering the alumina and ferric oxide on a second filter. The amount of silica, however, is not sufficiently large to be of practical importance.

The filtrate containing the lime was heated to incipient boiling and precipitated with an excess of $(NH_4)_2C_2O_4$, allowed to stand over night and next day filtered off and ignited over a blast lamp to a constant weight. The magnesia was precipitated with $(NH_4)_2HPO_4$ and ignited and weighed as pyrophosphate. The filtrate from the magnesium was concentrated to about 400 c.c., rendered slightly acid with HCl, and while hot precipitated with barium chloride and the barium sulphate determined as usual.

This scheme admits of the determination of the:

Insoluble residue,
Soluble silica,
Alumina and ferric oxides,
Lime (CaO),
Magnesia (MgO),
Sulphuric oxide (SO_3).

It was found that the detemination of the carbonic acid involved a large amount of work of questionable, practical utility, as, after a number of very careful determinations, it was found that the carbonic acid was in every case a small percentage less than the amount volatile at a red heat. This latter determination was made by heating one gram in a small covered platinum crucible, at a red heat, to a constant weight.

Carbon was found to be present in several instances in appreciable amount. It was determined by passing the original acid solution through a balanced filter and deducting the weight of the residue after ignition from the weight of the dried residue upon the filter.

Hydrogen sulphide was observed in a few instances escaping when the cement was dissolved in dilute acid.

The percentage of alumina and lime in any given cement soluble in water was found to be a varying and significant factor. Five grams were in two instances treated with 200 c.c. of distilled water and immediately filtered. The filtrate was acidulated with HCl, this rendered alkaline with NH_4HO, and after boiling, the precipitated alumina was filtered off. The alumina was found to be 0.15 per cent. and 0.00 per cent. The lime was then precipitated from the boiling solution with $(NH_4)_2C_2O_4$. In the two instances under consideration, the percentages of lime were found to be 1.696 and 1.694. This led to further investigation and the conclusion that the 200 c.c. of distilled water were in both instances saturated with CaH_2O_2. Finally 1 gram of cement was treated by bringing it into 500 c.c. of distilled water and filtering immediately after thorough stirring. The amount of lime was found to vary in the different samples from 1.18 per cent. to 8.18 per cent. The alumina dissolved varied from 0.00 per cent. to 1.11 per cent.

As soon as the cements were analyzed according to this scheme the results were arranged with the physical tests in several series, from lowest to highest, in order that any correspondence between the physical tests and chemical composition might be more apparent. No such correspondence was revealed in any case, and the results were apparently without relation.

On arranging the series with reference to the content of lime, from low to high, No. 49 appeared as second in the list, when in the series arranged upon seven days mortar tests it was nearly the last in the list and consequently one of the best. No reason was apparent for this striking lack of correspondence, and the analysis was repeated with carefully diluted 20 per cent. acid, into which the cement was slowly jarred. The result appears in the table given below as No. II., the first mentioned analysis being No. I. Another analysis was made by pouring concentrated acid upon the cement, with results given below as No. III. Still another was made by pouring upon the cement 10 per cent. acid, with still other results given as No. IV. A study of these results led to the suggestion that possibly the active chemism attending the solution in the several instances observed might be responsible for the varying results.

Five grams were then sifted through a No. 40 sieve upon the surface of 250 c. c. of 10 per cent. acid in an 8-inch evaporating dish. The mixture was then vigorously stirred at intervals for half-an-hour, or until all that would dissolve was in solution, when the solution was filtered and the residue on the filter thoroughly washed with hot water. The results obtained are given as No. V. Assuming that the magnesia, sulphuric oxide and loss at a red heat were correctly estimated in No. I., Nos. IV. and V. give, respectively, a total of 99.20 per cent. and 99.21 per cent. There is very little difference in the percentages of lime, but the sieve method yields the largest percentage of soluble silica. It was inferred from this result that perhaps a still weaker acid might dissolve more silica, and the attempt was made to decompose the cement with a 5 per cent. acid, but without success.

Number.	Insoluble in HCl.	Soluble Silica.	Alumina and Ferric Oxide.	Lime.	Magnesia.	Sulphuric Oxide.	Loss at Red Heat.	Lime Soluble in H₂O.	Alumina Soluble in H₂O.
I	20.42	12.24	7.50	55.05	2.19	1.90	1.15	5.27	0.52
II	7.87	16.37	8.88	62.32	2.19	1.90	1.15	5.27	0.52
III	22.39	1.63	8.98	62.74	2.19	1.90	1.15	....	....
IV	5.83	17.37	9.00	61.76	2.19	1.90	1.15	5.27	0.52
V	5.79	17.81	8.65	61.72	2.19	1.90	1.15	....	0.52

The series of Portland and Natural Cements were then analyzed with ten per cent. acid, using a No. 40 sieve to throw them upon the acid in such a manner as to reduce the evolution of heat to the lowest terms. The results were arranged on the seven-day mortar test, from lowest to highest as is shown in Table No. 3.

Meaning of Terms.

There are several terms used in this table that are here used, it is believed, for the first time. The required tensile strength in pounds for Portland cement is: One day, 160 pounds; seven days, 350 pounds, and seven days mortar, 125 pounds—the sum of which is 635 pounds. In computing the "total tensile strength in pounds," the tensile strength obtained in one day and seven days neat and seven days mortar tests are added together and are then made a means of comparison. By this means of comparison No. 149 is found to be of more than twice the required strength.

The terms "active" and "inactive" constituents represent, respectively, the active and inert portions of any given cement. The percentage of active constituents may be obtained by adding together the percentages of soluble silica, alumina and iron and lime. That of the inactive constituents by adding together the percentages of matter "insoluble in ten per cent. HCl," of "carbon," when present, of "magnesia," of "sulphuric oxide" and of "matter" volatile at a red heat.

The constitution of cements has been under discussion for many years, but no results that are final may be said to have been reached until those described by the Messrs. Newberry were published in 1897. It is not necessary here to review the work of Le Chatelier and Vicat. That work has been sufficiently well done by the Messrs. Newberry in their several papers. We wish, however, to re-state the Newberrys' results in order to point out their relation to our own work.

The Newberrys have shown, by the most elaborate and conclusive physical and chemical tests, that the formula for cements is "% lime = 2.8 (% silica) + 1.1 (% alumina)." They have further shown that in the manufacture of cement, alumina and iron may be taken together, while magnesia is inert and sulphuric oxide and the alkalies in small quantity may be disregarded. From all of which it is fair to conclude that the lime, soluble silica, iron and alumina are the constituents of a cement that give it value, while all the other constituents that it may contain are inert or injurious.

It may therefore be assumed that a perfect theoretical cement, formed synthetically from pure materials, would be found upon analysis to be represented by the formula,

$$\frac{CaO}{2.8 \text{ (soluble silica)} + 1.1 \text{ (alumina + iron oxide)}} = 1$$

One, then, becomes the index of the active ingredients of a theoretical cement, and the results of analysis may be used to express under the above formula the degree of approximation obtained in the manufacture of any given cement to that theory. This is the significance of the term "active index."

The "inactive index" is obtained by adding together the percentages of the inactive constituents with which the cement proper is diluted.

As the Portland and Rosendale Cements are wholly unlike substances they will be discussed separately. An examination of this table and comparison of the same as to Portland Cements with the table of Physical Tests shows, that the first five numbers are the same in each table, though differently arranged. Four of these five numbers are below the required tensile strength on seven days neat and seven days mortar tests, the fifth, No. 55, is below on one day neat and seven days mortar and barely up to the requirement on seven days neat test. On further examination it is observed that No. 58 heads all three lists. This is a poor cement all through, being low in lime, high in magnesia, with nearly one per cent. of carbon, and showing by the loss at the red heat of 28.86 per cent. either the effects of ageing or bad or insufficient burning. It is a very bad cement.

No. 55 is a cement high in soluble silica but low in alumina and iron. It dissolved in dilute HCl with effervescence leaving a residue consisting largely of anhydrous ferric oxide. This indicates a cement not properly balanced and badly burned. It is a poor cement.

No. 57 is not properly balanced. There is too much insoluble material, and nearly 2 per cent. of carbon, with a copious evolution of hydrogen sulphide on solution in dilute HCl. With concentrated HCl it completely gelatinized, with elimination of CO_2 and H_2S. There is not enough lime and soluble silica for the alumina. The high inactive index of 18.35 per cent. with 0.5 per cent. undetermined shows that the substance is really only about four-fifths cement.

Nos. 65 and 66 are too high in soluble silica and too low in alumina and iron, with too much magnesia.

No. 68 is a fairly good cement, but it contains too little alumina, too much magnesia, and is the only one in the list that contains so much sulphuric oxide as to lead to the suspicion that it has been adulterated with gypsum.

Nos. 71 and 67 are also fairly good cements.

Nos. 30 and 53 are two different samples of the same brand. Although they are above the required tensile strength they are poor cements, either from being underburned or from ageing. As they both contain carbon they can reasonably be supposed to be underburned. This results in an inactive index in No. 30 of 24.22, and in No. 53

of 18.39. So much inert matter, chiefly carbonate of lime, destroys the balance of the active constituents, and while the total lime that enters into the active constituents may be sufficient to give a proper active index there is only a part of such lime in such a condition as to properly combine with the silica and alumina and form the hydraulic mixture belonging to them—in other words, a part of the lime is in the form of carbonate of lime, and properly belongs in that condition to the inactive instead of the active constituents. The carbonic acid in No. 30 was found to be 12 per cent., which is equivalent to 27.276 per cent. of carbonate of lime. The 15.276 per cent. of lime thus found to be inert, leaves only 42.884 per cent. in combination with the silica and alumina to form about 61 per cent. of the mixture. This explains why these cements are so high on seven days neat, and so low on seven days mortar tests.

The remaining seven cements are all, including the slag cement, of very superior quality, and these results demonstrate to what a high degree of excellence the cement industry has attained in the United States. Of seventeen samples twelve are above test, and of the twelve eight are superior to the best foreign cements upon the market.

From the average composition of the seven best cements it can be seen how nearly the manufacturers of American Portland Cements approach a theoretically perfect cement. The tensile strength is nearly double that which is required, and in all higher than the best brands of imported cement. The soluble silica is 18.45 per cent., the alumina and ferric oxide is 9.46 per cent., the lime is 61.89 per cent. The active index is .997. The insoluble material is less than 5 per cent.; only a trace of carbon appeared in two of the seven specimens, with less than 2 per cent. each of magnesia, sulphuric oxide and matter volatile at a red heat—that is to say, that the burning is nearly perfect. All of these percentages give an inactive index of less than 10 per cent., which means that the seven best cements examined averaged more than 90 per cent. pure, and that this 90 per cent. is almost a theoretically perfect cement.

We think the manufacturers of American Portland Cement are to be congratulated.

ROSENDALE (NATURAL) CEMENTS.

It could not be expected that a cement made by burning a natural lime rock containing clay and silica would be found, on chemical analysis, to conform in its composition to any theoretical formula. The fact that natural cements do not could not be more forcibly demonstrated than by an inspection of the accompanying table. In general they are badly balanced according to Newberry's formula, showing an excess of lime over that required for the acid elements, silica, alumina and iron. In general also they exhibited the effects of both under and over burning. They nearly all contained ferric oxide, rendered inactive and insoluble, even in concentrated HCl. This condition of the ferric oxide is due to overburning. In general they effervesced on solution in dilute HCl and lost from 6 per cent. to 15 per cent. of matter volatile at a red heat. This reaction indicates from 12 per cent. to 30 per cent. of unburned limestone. No other results could be expected from careless and rapid burning of stone in large lumps in kilns or stacks than great lack of uniformity in the result and consequent uncertainty as to the value of the product.

At the price at which Portland Cements of the finest quality are now placed upon the market very little inducement can be offered for the use of Natural Cements in constructions where permanence and solidity are the first considerations.

Conclusions.

The facts hereinbefore stated lead to a number of conclusions of permanent value. The literature relating to Portland Cements contains references to the supreme importance of fine grinding too numerous to mention. A great number of experimenters have established the fact that all of the material contained in a Portland Cement that is rejected by a sieve of 100 meshes to the linear inch possesses very little value as cement. It was observed, when jarring the different specimens through a No. 40 sieve, merely for the purpose of distributing the cement evenly and slowly upon the surface of the dilute acid, that all of those cements that contained CO_2 effervesced more freely as the coarser particles were reached, towards the end of the operation. This indicates that the cement is more easily ground than the unburned carbonate of lime, and that consequently the cement contained in the mixture is really found in the finer particles while the coarser particles are not cement at all. This fact is particularly noticeable in the few cements that contain a notable quantity of sand, either fine ground or in grains of appreciable size.

It therefore occurred to us that a proper analysis of a Portland Cement should be made only of that portion that passed a sieve of 100 meshes to the linear inch. Several analyses were made in this way, but the results obtained indicated such radical differences from all recorded analyses that the plan was abandoned for the one herein described.

Mr. Shirmer, in his article published in the March number of the Journal of the American Chemical Society for 1899, advocates the elimination of silica from the portion of cement insoluble in acids, by treatment of these residues with hydrofluoric acid. His contention is perfectly sound, that this insoluble residue should not be determined as silica, as it contains lime, alumina and perhaps magnesia. Nevertheless, it is an insoluble residue, and the silica, lime, alumina and magnesia contained therein bear no relation to the silica, lime, alumina and magnesia that pass into solution. If the analysis is properly conducted it is of no consequence as determining the value of the cement, of what this residue consists, so long as it is not soluble in dilute acid, and is not in excessive amount. In the cements that we have examined this residue usually contains siliceous sand, either fine or coarse. In one Portland Cement it consisted largely of anhydrous ferric oxide. The cement was under test and probably overburned in part. In the natural cements this residue consisted in most cases of sand, ferric oxide and underburned rock, indicating a very uneven burning.

Many of the papers which we have examined are interesting as treating subjects for scientific discussion or information, but do not impress us as of any direct practical value. Much of the analytical work set forth or explained in them appears to us to be ultimate rather than proximate, and as such fails to exhibit those differences of composition in detail upon which the differences in value observed among cements may be supposed to depend.

It is not proper, in our judgment, that a cement should be dried either at a low or high temperature before analysis and therefore deprived of its water or carbonic acid, or both. The sample should be analyzed precisely as it is brought to the laboratory. The active constituents of a cement are all soluble in 10 per cent., HCl, when they are properly brought together. They are soluble silica, lime, alumina and iron. By no method of treatment should either of these be increased or diminished. In illustration of these statements let us further discuss Nos. 30 and 53. The total tensile strength indicates a cement from 20 per cent. to 50 per cent. above that which is required, as they are both different specimens of the same brand. The active index indicates a considerable excess of lime, but if this large excess had been active lime the cement would have been weaker and below test. There is a large amount of matter volatile at a red heat, some of which would have been expelled and lost in drying. Analysis showed 12 per cent. of CO_2 in No. 30, equivalent to 27.276 per cent. of carbonate of lime or unburned limestone, containing 15.276 per cent. of lime. When this lime is subtracted from the total lime—58.16 per cent.—there remains only 42.884 per cent. in combination with the silica, alumina and iron, giving an active index of 1.076 and also giving 61 per cent. of cement diluted with 39 per cent. of carbonate of lime and other inert matter. Such a cement would be expected to stand fairly well on a seven days' neat test and to nearly or quite fail on a seven days' mortar test. The mixture is 61 to 39 to begin with, and when 300 more parts of sand are added it becomes 61 to 339, or 1 to more than 5. Such a problem as this is not solved by ultimate analysis, with the determination of insoluble silica, lime and alumina.

Before closing this discussion we wish to express our unqualified appreciation of the researches carried on by the Messrs. Newberry and lately published in the journal of this society, Volume 16. It is difficult to place too high a value upon researches of this character. They exhibit an absolute mastery of the problem, and proceed to conclusions that within certain limits are final. These limits are included within the more or less complete application of methods of verification to which the results are subjected and upon which the conclusions are based. That is to say, if the results of a perfect theoretical synthesis are tested by perfect physical apparatus, confirmed by accurate chemical analysis, such tests would furnish an infallible guide in the practical manipulation of crude materials; provided that such practical manipulation is always as perfect as the synthesis.

In the Messrs. Newberry's paper no details are lacking concerning the physical tests, while nothing is said concerning those of the chemical analysis. The steps that lead to the declaration, that so much of a clay of a given composition plus so much of a pure limestone are required are all clearly set forth; but the uncertain element that enters every day the practical problem is, to what extent is it possible to realize these theoretical conclusions in the working of a cement plant? As a reply to this question we point to the best seven American cements, and especially to No. 88.

It was to determine, first, how far the results of chemical analysis would confirm and explain those obtained by physical tests; and, second, how far both would sustain these general theoretical conclusions when both are applied to the cements on the market, that this research was undertaken. We trust our work has not been in vain.

Respectfully submitted,

OTTO H. KLEIN,
S. F. PECKHAM.

Tables.

No. 1.

No. 2.

No. 3.

Milton Keynes UK
Ingram Content Group UK Ltd.
UKHW021701201124
451457UK00007B/164